THE LIFE SCIENTIFIC
VIRUS HUNTERS

ANNA BUCKLEY

WITH A FOREWORD BY JIM AL-KHALILI

W&N
WEIDENFELD & NICOLSON

First published in Great Britain in 2021 by Weidenfeld & Nicolson
This paperback edition published in 2022 by Weidenfeld & Nicolson
an imprint of The Orion Publishing Group Ltd
Carmelite House, 50 Victoria Embankment
London EC4Y 0DZ

An Hachette UK Company

1 3 5 7 9 10 8 6 4 2

A CIP catalogue record for this book is
available from the British Library.

ISBN (Mass Market Paperback) 978 1 4746 0808 4
ISBN (eBook) 978 1 4746 0753 7

Typeset by Input Data Services Ltd, Somerset

Printed and bound in Great Britain by Clays Ltd, Elcograf S.p.A.

MIX
Paper from
responsible sources
FSC® C104740

www.orionbooks.co.uk
www.weidenfeldandnicolson.co.uk

Praise for *The Life Scientific*

'*The Life Scientific* does that clever thing, making a difficult subject more accessible not by talking down to us but by talking to scientists who are so passionate about their subject that they entice us in'   Kate Chisholm, *Spectator*

'Each week Jim Al-Khalili interviews brilliant scientists about their brilliant lives in research. It's inspiring because their achievements are huge'   *The Times*

'*The Life Scientific* ... transmutes the explanation of ideas into discovery. The listener always feels in the same room as the speakers. In showing non-scientists why science offers so many paths to discovery it has no equal'
Gillian Reynolds, *Telegraph*

'This excellent series specialises in guests who are both fascinating and admirable'
David Crawford, *Radio Times*

ANNA BUCKLEY is the series producer of *The Life Scientific* on BBC Radio 4. She has worked with the presenter Jim Al-Khalili since the programme was launched in 2011 and has produced nearly a hundred interviews with leading scientists, revealing the men and women behind the latest scientific discoveries. She has worked in the BBC Radio Science Unit for twenty years and lives in London with her husband, environmental consultant Mike Quint and two teenage daughters.

PROFESSOR JIM AL-KHALILI OBE FRS is a theoretical physicist, author and broadcaster based at the University of Surrey. He has written eleven books and has presented *The Life Scientific* since 2011. He is a recipient of the Royal Society Faraday Prize, the Institute of Physics Kelvin medal and the inaugural Stephen Hawking Medal for Science Communication.

Also in this series

The Life Scientific: Explorers
The Life Scientific: Inventors

To Eliza and Rosie, with love.
May you each find your own way to transform lives.

CONTENTS

FOREWORD

This is the third book in this fabulous series based on interviews recorded for the BBC Radio 4 programme and podcast *The Life Scientific*, in which I interview leading scientists and engineers about their lives and work.

I find it hard to believe that we are now into our tenth year of broadcasting, and still seem to go from strength to strength. At the time of writing this Foreword, I have interviewed 225 guests and counting. Among them, nine Nobel Prize winners, four presidents of the Royal Society and three UK Government chief scientific advisors. And thanks to the seemingly endless supply of quite extraordinary scientists as well as all the fantastic work put in by my series producer, and author of this series of books, Anna Buckley, along with the other producers at the BBC Radio Science Unit, I fully expect the series to go on and on for as long as the universe continues to expand and doesn't collapse in a final Big Crunch.

Covid-19 has tragically touched all of us, whether directly or indirectly. These days, everyone fancies themselves as an expert on viruses and how they are transmitted. We all talk about R numbers, antibodies and quarantine and have opinions about how to stop the spread of infectious diseases. In this volume, you will read about the remarkable lives and work of scientists who really *are* experts in these fields, including two members of the Scientific Advisory Group for Emergencies, SAGE, that has been presenting evidence to the UK government during the pandemic.

Weaving together personal anecdotes and scientific insights, Anna Buckley has created a set of narratives about recent outbreaks and epidemics and what scientists have learnt from them that is essential reading in these extraordinary times.

Jim Al-Khalili,
October 2020

INTRODUCTION

This book was conceived during lockdown in the UK when millions of people were being held hostage at home by a microorganism that is one hundred thousand times smaller than the full stop at the end of this sentence. The British prime minister was in intensive care, having announced a few weeks before, on 12 March 2020, that 'many more families are going to lose loved ones before their time'. Hundreds of thousands of people had died worldwide; millions more were fearing for their lives. Our eyes had been opened to the threat of disease. We washed our hands more often and for twenty seconds and tried not to touch our faces, but confronted by a virus that had never been seen before, Homo sapiens (the wise one) suddenly seemed a little bit less all-knowing.

What was this virus and how did it spread? If only we could see these tiny particles suspended in the air between us and on surfaces all around. As we tried to make sense of this new threat, historians cast their minds back to the pandemics of the past. A lot was written about the Spanish Flu (1918–20) and the Great Elizabethan Plague (1665 and 1666), perhaps there was something to learn there. But scientists looked to the future and set to work. So much to discover, so little time. And I, like many others, became virus-obsessed. Non-stop speculation about 'the virus' was wearing. Uncertainty was inevitable. This coronavirus was new to humans after all. I wanted to know what was known and started thinking about who could be relied upon to tell us; several scientists sprang to mind. As a founder producer of *The Life Scientific* on BBC Radio 4, presented by Jim Al-Khalili, I had talked at length to men and women who had spent most of their lives studying viruses. These were the people I wanted to learn from and listening back to their *Life Scientific* interviews (recorded before Covid-19), I was shocked to discover how prescient they were. 'We need to realise that epidemics are not just things that go on far away that won't be relevant to us,' Jeremy Farrar had told Jim. Wendy Barclay had described how animal viruses can jump into humans and warned that when they do, they have the potential to cause pandemics. Kate Jones had warned that the threat from pathogens that live in wildlife was 'very significant and increasing'. The message from these scientists was clear: novel viruses pose a major threat to global health. It was not a case of if there would be a pandemic but when. The arrival of Covid-19 would not have been a big surprise to them. They had been expecting something like it for some time.

I remembered learning so much from these scientists at the time of the recordings but somehow, it had all felt less urgent back then. In a way that seems inexplicable now, their insights were of great interest to me but their warnings had fallen on deaf ears.

Now as we wait to see how bad this second wave will be, it seems all the more urgent to revisit the thoughts and experiences of some of the most eminent scientists in the field, the men and women who have studied viruses for decades.

Each chapter is based on an interview recorded for *The Life Scientific* on BBC Radio 4, with more information added for clarity and context. These are highly personal accounts told by scientists working at the forefront of discovery who travel to the centre of epidemics, treat patients in hospitals and work in laboratories. They present a sobering reminder of just how much was known about novel viruses and how they spread, long before the arrival of Covid-19, and provide an insight into how that knowledge was acquired.

When **Peter Piot** was a young medical student in the 1970s, he was advised against specialising in infectious diseases. Everything was under control, he was told. There was nothing more to learn. But 'being slightly stubborn', he went ahead anyway. Flying to Zaire in 1976 and being a junior member on the team that discovered the Ebola virus sealed the deal on his choice of career. Later, he investigated an epidemic in central Africa and showed that it was caused by the same virus that was killing gay men in Paris. It took him years, however, to convince the scientific community that HIV AIDS was a heterosexual disease. As the founding Executive Director of UNAIDS (1995–2008), Peter learnt the hard way – and much to his dismay – that if he wanted to stop the spread of HIV AIDS, he had to 'go political'. The scientific evidence would not speak for itself, he realised. He is now the Director of the London School of Hygiene and Tropical Medicine and in May 2020 was appointed as a special advisor on Covid-19 to the European Commission.

Witnessing young men dying from HIV AIDS within weeks of getting a diagnosis made a profound impression on **Jeremy Farrar**. He was working as a junior doctor in a hospital in London in the early days of the epidemic when there was no treatment for this novel virus. In his thirties, he applied 'out of the blue' for a job at the Hospital of Tropical Diseases in Ho Chi Minh City and worked there for eighteen years, as a doctor and medical researcher. He describes how Vietnam, a country with limited resources, was able to contain an outbreak of SARS in 2003. The prompt introduction of robust public health measures was key. The following year, together with his colleague Tran Tin Hien, Jeremy diagnosed the first case of H5N1 bird flu in the country and acted fast to stop it spreading. He became the Director of the Wellcome Trust in 2008 and sits on the Scientific Advisory Group for Emergencies (SAGE). Jeremy has been presenting scientific evidence to the UK government during the current pandemic, as has **Wendy Barclay,** who is also a member of SAGE.

As a young woman, Wendy had no particular desire to save the world from diseases, but she was curious. The apparent simplicity of Rhinovirus 14 appealed and she thought it would be interesting to find out how it worked. Her early research at the Common Cold Unit near Salisbury involved taking nasal swabs from hundreds of snotty humans who had happily volunteered for an all-expenses paid holiday with a one-in-three chance of catching a cold. The first coronavirus was identified at the Common Cold Unit in 1965 and was so-named because it looked as if it was surrounded by a crown. When the HIV AIDS epidemic took hold, finding a cure for the common cold felt less pressing and the unit was closed in 1990. Wendy went on to study the molecular machinery of influenza viruses using pioneering techniques to sequence their genomes, seeking to understand why some animal viruses are more deadly to humans than others. Wendy and her team's analysis of the much-feared H5N1 bird flu virus showed that it was 'really quite hard' for this particular virus to infect human cells. Not all the news in virology is bad. More recently, Wendy's team at Imperial College in London showed that SARS-CoV-2 can be transmitted in tiny droplets in the air known as aerosols – a key bit of evidence in support of the idea of wearing face masks.

Jonathan Ball is the son of a coal miner. He might have followed in his father's footsteps but, encouraged by his teachers, he stayed on at school and went to Bristol Polytechnic. His first job, as a research assistant at a high security laboratory in Porton Down, involved manufacturing HIV in the laboratory so that it could be studied and understood. He hunted for different strains of HIV within our bodies and pioneered the study of this virus in the genital tract figuring that these viruses were the ones that needed to be disabled if transmission was going to be stopped. He has worked on antibody-based treatments for HIV and Hepatitis C and now, together with his team at Nottingham University, he is studying the antibody response to Covid-19.

Rather than focus on individual pathogens, ecologist **Kate Jones** is interested in where outbreaks of disease are most likely to occur. She created a map of disease hotspots around the world and showed that areas on the edge of forests are particularly vulnerable to outbreaks of disease. Deforestation creates more opportunities for the viruses that live in wildlife (often bats) to spill over into humans. Typically they find a way to infect our domesticated animals, such as chickens or pigs, and then us. We can blame 'the virus'. We can blame the animals that host it or our politicians for failing to take action. We can make scapegoats of scientists, political advisors and super spreaders. But if we fail to appreciate how we may have contributed to these crises, then we are missing an important part of the equation. The best way to prevent the next pandemic, Kate and many others believe, is to respect wildlife and protect wild places. When the Covid-19 pandemic was announced, she tweeted: 'Don't blame bats for coronavirus, blame humans.'

Martha Clokie sees viruses in a different light – as an opportunity, not a threat. There are trillions of viruses on planet earth and not all of them are out to get us. In fact, the viruses that infect bacteria (bacteriophages) could be used to cure disease. These are the viruses that Martha wants to find. Her research career began studying the evolution of African violets in Uganda. Or perhaps even earlier, collecting seaweed on Scottish beaches as a child. She was, and still is, passionate about plants but mid-career – and quite by chance – she met some scientists from Georgia who had been using viruses to treat gut and skin infections and decided to turn her attention to medical research. She is perhaps the most committed virus hunter in this volume, having spent nearly a decade searching for the viruses that are associated with C. difficile, bacteria that cause a particularly nasty, and often deadly, form of diarrhoea. Eventually, she found what she was looking for buried in a salt marsh. Martha's work offers us hope that viruses could be used to improve our health. Viral medicines could provide much-needed alternatives to antibiotics. When the pandemic hit, Martha started looking for viruses that could be used to treat the bacterial lung infections that can result from Covid-19.

}

I have been informed, educated and – yes, at times – entertained by these wonderfully honest accounts from scientists working on the frontline against viruses that make us ill and, in Martha's case, hunting down viruses that might cure not kill. I find it inspiring and reassuring to learn about the twists and turns in their careers and hear about moments of insight and despair. I hope this collection of scientific life stories will highlight the quiet but valuable contribution made by all the men and women (not just those featured in this volume) who have worked so hard for so many years without much notice or encouragement from the rest of us. Their work made it possible for scientists around the world to respond with astonishing speed when people started dying from a mysterious new illness in the city of Wuhan in December 2019. Without their accumulated wisdom, skills and expertise, we would not have known where to begin.

PETER PIOT

Stopping the spread of HIV AIDS

Born: 1949

Grew up in: a small village in Flanders, Belgium

Occupation: doctor and microbiologist

Job title: Director of the London School of Hygiene and Tropical Medicine

Viruses studied: Ebola, HIV

Inspiration: 'staring at microbes down a microscope and later seeing the suffering of people with HIV'

Passion: 'talking to people at the centre of an epidemic'

Mission: 'to make sure that everyone has access to the anti-viral medicines and vaccines they need'

Advice to young scientists: 'follow your passion and be the best'

PROFESSOR PETER PIOT has spent much of his life in pursuit of deadly viruses. He was one of the first people in the world to see the Ebola virus under the microscope. When wealthy Africans started arriving in Belgium in a desperate state, Peter feared he was witnessing a tiny fraction of much greater suffering and travelled to the Democratic Republic of the Congo (then Zaire) to investigate. It took him years to convince fellow scientists and health officials that the epidemic he uncovered was caused by the same virus that was killing gay men in Europe and the USA. In the early days of this epidemic, no one believed that HIV AIDS was a heterosexual disease. Mid-career he realised that if he wanted to stop this epidemic he needed to start thinking less like a scientist and more like a politician. As Executive Director of UNAIDS (1995–2008) he ensured that antiretroviral treatments were made available to everyone, not just those who could afford them. In 2010, he moved to London and became Director of the London School of Hygiene & Tropical Medicine.

Jim interviewed Peter in February 2016 in the wake of an Ebola epidemic in West Africa that claimed more than 11,000 lives.

The advice Peter Piot was given in his final year of medical school at the University of Ghent was clear: 'There is no future in infectious diseases. Don't waste your time on this. We have vaccines. We have antibiotics. And we have good sanitation.'

'They really thought there was no threat from infectious diseases?' Jim asked, aghast.

'Yes. They thought it was all over,' Peter confirmed.

Pandemics were thought to be a thing of the past. Spanish Flu belonged in the history books. 'That was really common wisdom in the 1970s and early 1980s, until, you know, HIV AIDS hit . . . Who would have thought that at the end of the twentieth century a new virus would pop up that, cumulatively, has now infected 80 million people and killed more than 30 million?'

The medical consensus in the 1970s was that viruses were under control. Smallpox had been eradicated from North America and Europe in the 1950s and a global eradication programme to rid the world of this much-feared disease was making excellent progress. Polio cases were falling rapidly following the introduction in 1962 of an oral vaccine, administered in drops or on a sugar cube. We had vaccines for measles, mumps and rubella. HIV, Ebola, Nipah, Zika, Swine Flu and the coronaviruses that cause Severe Acute Respiratory Syndrome (SARS), Middle Eastern Respiratory Syndrome (MERS) and Covid-19 were not yet known to humans.

'Better to work on cancer . . .' Peter was told, when he expressed an interest in microbes, epidemics and helping developing countries. But 'being slightly stubborn', he stuck to his guns and got a job as a microbiologist at the Institute of Tropical Medicine in Antwerp.

The work at the Institute was unglamorous and routine. Staring down a microscope at stool samples received from Belgians with unspecified gastro-intestinal disorders, Peter didn't feel as if he was saving the world, but one grey day in October 1976, a parcel arrived from Yambuku, an isolated village in the heart of the Democratic Republic of the Congo (then Zaire). Peter watched, excited, as his boss opened it up and extracted two vials of blood from a battered blue thermos full of melted ice. The blood samples belonged to a Flemish nun who had died in the most horrible way, bleeding from every orifice and with blood blisters bubbling up under her skin.

Professor Jean-Jacques Muyembe and the local doctors suspected yellow fever – symptoms like these were rare but not unheard of – but when the team isolated a virus from the nun's blood and studied it under an electron microscope, it looked nothing like the Flavivirus that causes yellow fever. Indeed, it was quite unlike anything Peter had seen before. Most viruses are spherical or squares. This one had a strange worm-like structure and, compared to most other viruses, it was huge. When Peter consulted *An Atlas of Viruses*, a hefty volume first published in 1972 with detailed diagrams of all the viruses that were known at the time, the only worm-like virus he could find was the Marburg virus.

At this point, Peter's mind started racing. Very little was known about this virus. It had been discovered in Marburg, Germany nine years earlier, when a group of workers in a pharmaceutical company had died suddenly, after having been in close contact with cell lines taken from a group of monkeys from Uganda. Had there been another outbreak of this deadly virus in the Democratic Republic of the Congo? If so, this would be the research opportunity of a lifetime. Peter pleaded with his boss to be allowed to continue work but was told that a Marburg-like virus was too dangerous to be handled in a regular laboratory in Antwerp.

Following the protocol for viruses of such a deadly nature, Peter's boss alerted the World Health Organisation and was promptly ordered to send the samples to the Center for Disease Control (CDC) in Atlanta, USA.

Bitterly disappointed, Peter returned to his routine work.

A week later, he was on a plane to Kinshasa, DRC.

'I had never been to Africa. I had never investigated an epidemic. I was not qualified in any way. But I wanted to go. And there were not that many volunteers!'

His colleagues cobbled together some safety equipment for him – scrubs, gloves and motorcycle goggles – and Peter read all the scientific papers he could find on haemorrhagic fevers, like Marburg. There were not very many.

'So you had gone straight from the lab, looking down a microscope at these samples, to joining a team over in Zaire?' Jim asked.

'Yes. And it was a fantastic adventure!' Peter said, enthused. 'I discovered parts of Africa, and I loved it. And I discovered the power of international collaboration. Science is one of the most globalised activities in the world today; a 24-hour cycle, with teams that span the globe.'

While Peter and the team travelled to the source of this unknown virus, the convent of The Sisters of the Sacred Heart of S'Gravenwezel in Yambuku, DRC, the CDC scientists in Atlanta discovered that the virus Peter had sent them was not the Marburg virus. And in a separate development, scientists in the UK had identified a new Marburg-like virus that was responsible for an outbreak of haemorrhagic fever in southern Sudan.

When the scientists arrived in Yambuku, the nuns warned them to stay away: 'To enter is to die!' one screamed. All around people were immobilised by violent vomiting and chronic diarrhoea and were bleeding to death.

'I am trying to put myself in your shoes,' Jim said. 'You have been wrenched away from the safe, controlled environment in Antwerp and plunged into this very different, very distressing, very dangerous, environment. Weren't you afraid that you would die?'

'I got scared when it was all over. But at the time, it was the excitement of the discovery, discovery of the new virus, discovery of an epidemic, discovery of – I would say – a continent of people. Discovery of myself, that, yes, I can do this.'

Surrounded by suffering and death, Peter felt alive. Away from home and the strict norms that had governed his childhood, he discovered skills he didn't know he had. 'I am from a fairly traditional Flemish background, a small village, and my mother would say all the time: speaking is silver, silence is gold, you have to work. There is this very strong work ethic and you never complain, you just do your job and you are not educated to become, let's say, a leader. I went to a regular local school. Not a school for leaders or anything like that. And then suddenly I found: yes, I can do much more, I can lead a team, I can investigate things.'

He sat down with the villagers and asked them what had been going on and, within a week, the team had worked out the main transmission routes for this mystery Marburg-like virus. 'One, we found that caring for someone who was infected, having close contact with them and being exposed to their body fluids, that is how you get infected,' Peter said. 'But, secondly, we found out that every morning the Mother Superior of this convent, a Catholic convent, would distribute five needles and syringes to the Outpatient Department.' The same needle was used to inject dozens of people every day and, in between each patient, 'the most they would do was flush the needle a little bit'. Tragically, the nuns' attempts to inoculate people to protect them from other diseases were the best possible way to transmit this terrifying virus which, as scientists later learnt, cannot survive very long without being surrounded by bodily fluid.

Viruses are often named after the places where they are found but Karl Johnson, the CDC team leader, thought this was a bad idea. He felt sorry for the residents of Marburg, Germany. The people of Yambuku had suffered enough, he felt, without the added burden of a deadly reputation, and the team decided to name this virus after a nearby river. Ebola.

'You are known as Mr Ebola and with good reason,' Jim said. 'But you have spent much longer – two decades – working on HIV AIDS. When did you first witness the devastating impact of this scary new syndrome?'

'In the early 1980s. I was still working at the Institute of Tropical Medicine in Antwerp. We were seeing more and more patients in our hospital coming from central Africa, both Africans and European ex-patriates who were living there, with a syndrome, with a disease. We didn't know what it was. And they all died ... We had never seen anything like that – neither me (I was still young), nor more experienced clinicians. These patients died so quickly. Whatever we tried in terms of treatment failed. We had never seen this happening in fairly young adults.'

If the situation in Antwerp was worrying, what was happening in the countries that these people had left? Peter wondered. He was seeing the wealthy elite – the politicians, business people and expats – the people who had flown to Antwerp for their medical care. What about all the people in central Africa who couldn't afford to hop on a plane?

'I try to go and see first-hand what is going on. I go into the lion's den'

'So I did what I do,' Peter said. 'If I think there is a problem, I try to go and see first-hand what is going on. I go into the lion's den.'

As soon as he arrived at Mama Yemo hospital in Kinshasa, he knew it was serious. The hospital was overrun. Men and women were lying in the open walkways, emaciated and desperate. It didn't take long to work out that this was a major epidemic.

'I took a deep breath,' Peter said. 'And thought to myself: "This is a catastrophe. This is a disaster. This is going to spread all over Africa ... This is what I am going to work on. And I am not going to rest until we can stop this epidemic." '

The team had managed to contain the outbreak of Ebola in Yambuku and Peter intended to do the same again. But he admitted to Jim, 'I totally underestimated how difficult that would be.'

}

As the full horror of the situation was sinking in, Peter wondered what *this* new virus could be? He had seen patients in Antwerp who were in a similar terrible condition. They were all gay men. All sorts of men and women were suffering in the epidemic in Africa so it was clearly not 'a gay syndrome' as some people had called it. 'There were some different clinical manifestations. The type of lung infections were different. There was far more tuberculosis.' But the progression of the disease – the way it seemed to be attacking the immune system – was remarkably similar, Peter thought.

To test this hypothesis, he wanted to see if patients who had been infected in Africa had developed the same antibodies as the people who had been infected in the West. And so he wrote to Luc Montagnier,[1] at Institut Pasteur in Paris, to see if he might be able to help him in his experiment. He didn't expect to get a reply from such a leading light in the emerging field of HIV research, but to his great surprise, Luc wrote back saying he would be happy to help. Delighted and excited, Peter sent him two sets of blood samples just as soon as he could. One

1 Luc Montagnier discovered a new virus – which he called LAV – in a biopsy of a patient with Acquired Immune Deficiency Syndrome (AIDS) in 1983. Together with Françoise Barré-Sinoussi, he was later awarded a Nobel Prize for the discovery of HIV, the cause of AIDS.

set was made up of blood taken from patients in Mama Yemo Hospital who were dying from this new disease. The other was taken from people in the community. This was the control. And to avoid any human biases creeping in, he made sure that the testers in Paris had no way of knowing which was which, mindful of the fact that scientists are more likely to find what they are expecting to see.

'One day in February, Luc called me,' Peter said. 'That was an incredible moment!' Luc read out his results over the phone. 'Number 1 negative, number 2 positive, etc.' And Peter, who had his code telling him which sample was which in front of him, interpreted the results. 'I had the code numbers. I knew who we thought had AIDS and who had not.' The results couldn't have been clearer. 'All the cases that we thought were AIDS had antibodies against HIV, just like a gay man with an HIV infection in Paris . . . That was the ultimate confirmation.'

Later, Peter and the team were able to prove beyond all reasonable doubt that the virus that had infected men and women in Africa was the same as the virus that was killing gay men in Paris. They managed to isolate the virus from the blood sample of patients in Kinshasa and when they studied it in more detail and sequenced the genome, Peter's hypothesis was confirmed. It looked exactly the same as the virus HIV that had been isolated and found in patients in Europe.

Peter and his team were convinced, but getting the other scientists who were working on HIV to take their findings seriously was a major challenge. The idea that this was a 'gay syndrome' was so firmly lodged in their minds, that they just didn't believe Peter's results. 'Even our seminal observations of a big epidemic in central Africa was first turned down by the *Lancet*,' Peter said. 'Afterwards they published it and it was a very widely quoted paper . . . One reviewer in the *New England Journal of Medicine* said, "It is a well-known fact that AIDS cannot be transmitted from women to men." Again our paper was rejected.'

'These papers that you were trying to publish were ones where you were saying that the disease that you were seeing in Africa, was in fact HIV AIDS and they were saying it is not possible because women don't get the disease?' Jim said, checking that he had understood.

'Yes. It was really the scientific, or the AIDS community, the very small one in those days, that had a hard time accepting that this was not a gay disease. It illustrates for me that in science we are guided too much by convention. We should constantly question: what is conventional wisdom? That is certainly one of the lessons.'

> 'It was really the scientific, or the AIDS community, the very small one in those days, that had a hard time accepting that this was not a gay disease'

'So how long then did it take you to convince the world that HIV AIDS was a heterosexual disease?' Jim asked.

'After the initial observations in 1983, we needed to accumulate the evidence and so I think it took maybe two or three years. That was in the scientific community.'

It took another fifteen years, and a completely different strategy, to convince the people who had the power to make a difference that this was a major problem.

'And that was so frustrating,' Peter said. 'Around 1992, having worked on AIDS for about ten years, I had some kind of midlife crisis. It was very exciting doing research. We worked out the basic epidemiology, characterised the origin of the virus, understood that it came from chimpanzees and so on, and so on. But, every year, millions more people were becoming infected.'

For how long can I study this unfolding disaster? Peter wondered. 'It was very good for my scientific career,' he admitted. But it didn't feel right. 'I didn't only want to study the world. I wanted to change it.' He wanted to stop this terrible epidemic. 'That's when I had the maybe bizarre idea to take a sabbatical year and spend it working for the World Health Organisation. It is probably the worst place on earth to go if you want to change the world!' Peter joked. 'But on the other hand, I liked it. So I made a transition from being a scientist, and doing clinical work in Africa, to becoming a UN bureaucrat.'

> 'That's when I had the maybe bizarre idea to take a sabbatical year and spend it working for the World Health Organisation'

'This had never, ever been my career plan,' Peter stressed. 'I never had one, by the way!' His sabbatical *year* turned into a thirteen-year stint working for the United Nations. In 1995 Peter became the first Executive Director of UNAIDS.

'How easy was it for you, personally, to make that shift from believing that the important thing is to gather evidence, to make a scientific case, to making the political decisions, coming up with the political arguments and strategy to convince people?' Jim asked.

'The beginning was very difficult,' Peter replied. 'I didn't understand the environment ... I thought that it was self-evident. I thought that the numbers would speak for themselves. And we had solutions.' It was 'very frustrating' and 'yes, it was very upsetting. But, again, if it doesn't work the first time [you have another go]. And you never give up because the cause was far too important. What was at stake were millions of lives. The alternative – giving up – was not an option.'

'Did you ever feel overwhelmed by the sheer scale of the task, you mentioned millions of people continued to be infected, did you ever feel that this was just too much to cope with?' Jim asked.

'I was overwhelmed every single day,' Peter said. 'I would wake up and, on the one hand wonder what is the next crisis

'I was overwhelmed every single day'

of the day? Who knew that people's human rights could be violated in so many different ways? And then there was the infighting in the system, which was actually the worst.'

Peter hated the politics of it all. 'I was not there for a UN career,' he said. 'That never interested me. I was there to make a difference in the fight against AIDS.' He had risen through the ranks at the UN Headquarters in New York, but he didn't want to lose sight of the people who he was trying to help. 'It was really deliberate,' he explained. 'Wherever I would go I would sit down with groups of people living with HIV, and with AIDS activists.' Often, they would give him a hard time ... 'I must have a masochistic side maybe as well,' he said. It wasn't always easy talking to people who were living with the disease who were not getting the help they needed, but it was far more informative than reading official reports, or sitting in meetings with government officials and UN colleagues, Peter thought, and it reminded him that the work he was doing was about helping people, not writing reports.

}

It was an uphill struggle. 'After three-and-a-half years in charge of UNAIDS, I felt that we had gone nowhere. It felt as if we had made no progress. The number of deaths continued to increase. The number of new infections was increasing. There was no access to treatment. And so I did what I often do when I think we are stuck, I called a private meeting. I invited friends, yes. But also people who were very critical.' There was to be no record of the meeting. No notes were taken. 'And then I said: "OK, let me put it on the table, I feel like a total failure."'

'And then I said: "OK, let me put it on the table, I feel like a total failure"'

This honest assessment of the situation led Peter to an uncomfortable conclusion. Up to this point he had been proud of his scientific credentials and was determined 'not to go native'. He didn't want to be a bureaucrat. His strategy had been to provide the best possible scientific evidence, 'and then kind of hope that things would follow'. 'I felt that scientific insight was what gave me credibility. Why otherwise would I be different from any other politician?' But during the meeting, he was forced to admit (to himself as much as to everyone else) that the way to succeed was to stop thinking like a scientist. 'The conclusion of that meeting, which was a very tough one for me, was that we had to go political.'

The evidence would not speak for itself. 'It was all about the politics, which is not a negative thing. It's just the way to make things happen.' Instead of putting all his effort into presenting the best possible scientific evidence, Peter began to consider the politics of the situation. Who had the power to make things happen? Once he started thinking more like a politician and less like a scientist, things started to change. 'Let's put it this way,' he explained. 'It is the Security Council that really has power, it decides on war and peace ... with all due respect, it is not the Ministers of Health.'

A year later, on the 8th January 2000, HIV AIDS became the first health issue ever to be discussed at the UN Security Council. 'That opened so many doors,' Peter said, exasperated. 'It is completely stupid! Some Heads of State of countries who had literally millions of people living with HIV had consistently failed to take action. When it was discussed at the Security Council, they thought "oh, this must be serious".'

}

At a conference in Vancouver, in July 1996, it was announced that AIDS is treatable. 'It was a fantastic breakthrough ... thanks to basic research and investment by pharma.' Highly active antiretroviral treatment (HAART) 'completely changed how we look at AIDS'. When this treatment was incorporated into clinical practices in wealthy countries, it led to a 66 to 80 per cent reduction in the death rate from HIV for those who could afford it. Peter was determined to make these drugs accessible to those who need them most. 'And that was people in developing countries.' And so began another lengthy struggle.

'If you are motivated by quick results in medicine, you become a surgeon,' Peter said. 'In public health, in population health and epidemic control, you need a long-term goal. The goal was to stop the epidemic, to make sure that people were no longer becoming infected. It was also to ensure that those who are infected no longer die.'

The development of highly active antiretroviral treatment was 'a game changer' but the drugs were fantastically expensive. 'The price of antiretroviral treatment in 1996 was about $14,000 per person, per year.' And 'to bring down the price of these drugs was against common wisdom', he said. Even organisations that Peter thought would be allies seemed reluctant to help. 'I think there were more meetings organised by specialist international development and public health and health systems to discuss why it would *not* be possible to introduce these drugs [to the developing world]. And even, why it is bad to do this. Rather than asking, "OK, *how* are we going to do this?"'

This 'can't do, won't do' attitude infuriated Peter.

'It was not just frustrating. I was very angry. You know, governments did not want to work with this idea. I came here to the UK, for example, several times. The Department for International Development was dead against treatment for people with HIV in developing countries. UNICEF, WHO, the World Bank – all these people, these institutions, who were supposed to defend the interests and the health of people – they didn't want to do it.' In South Africa there was 'an extra complication because President Thabo Mbeki believed that HIV was not a cause of AIDS'. Misinformation like that, coming from the top, cost a lot of lives in South Africa and elsewhere.

'Attitudes are very different now, of course,' Peter said. But in the 1990s the political will was sadly lacking to make treatments affordable and available to people who were living with HIV in poorer countries. Eventually the price of antiretroviral drugs was brought down, thanks in no small part to Peter's lobbying. He continued to spend as much time as possible with people who were living with HIV all over the world, and used his knowledge of how the UN worked to make sure that they were represented at the highest possible level in UN policy discussions.

Having worked so hard and achieved so much to protect us all from HIV AIDS, Peter stepped down from his job as the Executive Director to UNAIDS in 2008. Jim asked him why he decided to quit.

'One, I think that presence for life is not a good idea,' Peter said. 'I mean you need new blood. Two, I felt that I had achieved what I could ... And I was very tired – literally, physically and mentally – and I wanted to do something else ... But I was also still very angry about the continuing resistance and discrimination against people with HIV AIDS. People who are, you know, in institutions or in power are not necessarily driven by what is best for the population. Often, they are driven by their own agendas. That said, I also discovered great politicians, people who move agendas, people who are courageous.'

> *'I was very tired – physically and mentally – and I wanted to do something else'*

'So much progress has been made and there is a lot of talk these days about eradicating AIDS by 2030. Is that a realistic aim?' Jim asked.

'I don't believe that it is possible without a vaccine,' Peter said. 'Slogans about eradicating HIV by 2030 are based on mathematical models that assume, for example, that everybody will take treatment, every single day for the rest of their life. But, you know, people are not robots and people do forget. There are stockouts. Budgets for HIV are in decline. And it is about sex, also. Let's take London. Here in London every single day, five gay men are becoming infected with HIV. In a country where we have free access to treatment through the NHS and there is a lot of testing. People know about HIV, but the perception is there now that, "it is all over and if I get it, I take a few pills".'

More than 80 million people are known to have been infected with HIV AIDS and it has claimed in excess of 30 million lives worldwide. And those numbers continue to rise. 'Statements about the end of AIDS are not very well supported by the reality of what's actually happening today,' Peter said. 'And I think it is actually dangerous because it leads to complacency.

'There are hugely complex scientific obstacles around finding a vaccine against HIV. What we can achieve is bringing down the number of new infections to much lower levels. There are now just under two million new infections a year, down from four million. That is fantastic progress. But that's still two million people getting infected and there are still over a million deaths a year.

'We can all make a difference. We have to find our way. We need to do better.'

Meantime the Ebola virus remains just as deadly today as it was when Peter first saw it under the electron microscope in Antwerp in 1976. 'Between 1976 and 2014 there were about twenty-five known outbreaks of Ebola in Central Africa. All limited in time, and in mostly rural areas, and with between 100 and 200 cases.' An outbreak in eastern Guinea in 2014 was different. It spread to neighbouring Liberia and then to Sierra Leone. Nine months into this epidemic, cases in the Liberian capital were doubling every two weeks. It was a perfect storm.

The virus infected people who were living in densely populated urban areas on the outskirts of cities, in closely packed corrugated iron shacks with poor sanitation and limited access to medical care. The healthcare systems in Liberia and Sierra Leone had been destroyed by years of civil war. And tragically, in many communities, funeral practices facilitated the spread of the disease. When relatives of the deceased lovingly washed the bodies of their nearest and dearest to prepare them for burial, their bodily fluids mixed creating an ideal opportunity for the Ebola virus – which can be transmitted through sweat, vomit and urine, as well as blood and sex – to infect its next victim.

This epidemic claimed more than 11,000 lives. It was on a scale that had not been seen before. As it came to an end, Peter warned: 'A large Ebola epidemic can and will happen again, unless we always remain extremely vigilant, respond promptly and have more to offer than isolation and quarantine.' The next time Ebola reared its ugly head, however, the world was better prepared.

When terrible memories of the West African epidemic were still fresh in our minds, the Global Alliance for Vaccines and Immunisation (a private public partnership between research institutions and pharmaceutical companies) provided funding to fast-track the development of an Ebola vaccine and a global emergency stockpile was created. At the first sign of an outbreak in the Democratic Republic of the Congo in 2018, 300,000 doses of this vaccine were made available for free, and operational assistance provided to launch a vaccination programme that helped to contain the outbreak.

'Are we getting better at dealing with these crises?' Jim asked.

'We have better technology which makes it easier to diagnose disease, and we have better communication within and between countries.' These two things make it easier to contain outbreaks and prevent epidemics from becoming pandemics. We have vaccines for Ebola but not for HIV. 'But anything can happen,' Peter said. 'New viruses pop up. And they all come from animals. HIV from chimpanzees, Ebola from bats, influenza from poultry, birds and pigs'.

There have been so many epidemics since Peter first chose to study infectious diseases as a young man, despite being warned that it would be a waste of time. 'Four decades on from the discovery of Ebola, I imagine you are a very different man?' Jim said.

'I'm not sure if I am all that different in terms of my values and what makes me tick,' Peter said. These days there are mathematical models to predict the likely spread of disease but, wherever possible, Peter remains committed to shoe-leather epidemiology. So much can be learnt 'from talking to people and just from being there'. He will still 'go into the lion's den'. 'But I have got more experience, of course,' Peter said. 'I am less naive about how the world functions and how you move agendas . . . I thought, forty years ago, that if you have got the evidence, if you have got the science, the rest will follow. But that, of course, is not how society works.'

JEREMY FARRAR

*Containing outbreaks of SARS and
bird flu in Vietnam*

Born: 1961

Grew up in: Singapore, New Zealand, Cyprus and Libya

Occupation: doctor

Job title: Director of the Wellcome Trust

Viruses studied: bird flu, SARS, Dengue, MERS, HIV

Inspiration: 'the joy of not knowing'

Passion: 'making discoveries and sharing that knowledge to help to save lives'

Mission: 'making the world a better place by reducing global inequality'

Advice to young scientists: 'dream a bit'

DR JEREMY FARRAR is a veteran from the frontline in the battle to contain outbreaks of infectious diseases in Southeast Asia. For eighteen years (1995–2013) he ran the Clinical Research Unit in Ho Chi Minh City, transforming it from a couple of rooms at the back of the Hospital of Tropical Diseases to a world-class bio-medical and public health research facility with more than 500 clinicians and scientists. In 2003, he witnessed the devastating impact of the SARS epidemic and saw how this novel coronavirus was contained by robust public health interventions. Together with his colleague, Tran Tinh Hien, Jeremy diagnosed the first human case of H5N1 bird flu in Vietnam and introduced measures to stop it from spreading. In 2011, he was awarded the Ho Chi Minh City Medal for his services to tropical medicine by the Vietnamese government. Eight years later (in 2019) he was knighted by the Queen for services to global health.

Jim talked to Jeremy in July 2014, nine months after he took up a new job in London as Director of The Wellcome Trust, a global charitable foundation, with an annual budget of £1 billion, that is committed to improving animal and human health.

Infectious diseases don't respect borders. 'What happens in Ho Chi Minh City today is relevant in London tomorrow. Or Washington, or Geneva, or Paris, or Jakarta,' Jeremy told

'What happens in Ho Chi Minh City today is relevant in London tomorrow'

Jim. 'And we need to realise that epidemics are not just things that go on far away, that won't be relevant to us.'

It's a message that seems utterly obvious to us all now. But Jeremy was talking to Jim six years before the coronavirus responsible for Covid-19 spread across the world in 2020, reminding everyone that, for microorganisms, there are no barriers between nations.

Our ever-increasing appetite for flying has created unprecedented opportunities for humans, *and* viruses, to travel around the globe. 'When I first came to the UK, I came by boat from New Zealand, to Southampton,' Jeremy said. 'It took eight weeks. If I had got on the boat with a nasty infection, acquired, let's say, on a stopover in Singapore or Egypt, I would have either died or recovered by the time I got to London. Now I can do that journey in twenty-odd hours.' And so can all the pathogens on board. High-speed international transport links have brought us all closer together. Globetrotting microbes spread infections and put us all at risk. 'The world is actually a very, very small place.'

Appropriately, perhaps, for someone who later specialised in infectious diseases, Jeremy grew up feeling that he was a citizen of the world. His father was an English teacher and his mother an artist and Jeremy and his five older siblings 'were dragged around by our parents, who left the UK in the early 1950s and finally came back in 1979'. He was born in Singapore and lived in Cyprus and New Zealand as a child. His teenage years were spent in Libya which he describes as 'a wonderful country to live in in the 1970s': a place of extraordinary natural beauty, generous warm people and with stunning Islamic architecture and culture and impressive Roman remains. Leptis Magna, the third most important city in the Roman empire is one of the best-preserved archaeological sites in in the world.

His first job aged 17 was filing visa applications in the British Embassy in Tripoli. Back then, he wanted to be a diplomat and was hoping to study Politics, Philosophy and Economics at university. But, after more than a term in the sixth form, he decided he wanted to become a GP and switched from studying English, Economics and History A levels to Biology, Chemistry and Maths. Midway through a medical degree at University College London, he took a year out to do a research project with the distinguished developmental biologist Cheryll Tickle, and his life plans changed once again.

'Cheryll was an absolutely brilliant, brilliant scientist,' Jeremy said. 'And an absolutely inspiring individual. Something of a throwback to the 1960s, actually. With rainbow-coloured trousers, always!' She was working on chick development. How does a single fertilised cell develop into a fully-fledged chick and, in particular, how do the chick's limbs and wings grow inside the egg. 'Something I have never looked at again,' Jeremy said, smiling. But it introduced him to the joy of scientific research.

'Medicine is a wonderful career ... But I have to admit I have never been brilliant at learning long lists of facts. Doing research ... you could dream a little bit beyond facts. You could ask questions, and

'*Doing research ... you could dream a little bit beyond facts*'

you could design experiments to try and answer them.'

Working with Cheryll was a world away from cramming for medical exams, learning and regurgitating what was known. It was creative and unpredictable. Jeremy relished the uncertainty of it all. 'It was that Cheryll Tickle experience,' Jeremy said laughing, 'that really persuaded me that I wanted to go into research.'

'How did you end up in Ho Chi Minh City?' Jim asked.

'Again, I would love to say that it was by design. But it wasn't. I was coming to the end of my PhD in neurology. I had sort of done my clinical training and ... I was giving a talk, in Norwich, to the Association of British Neurologists. In the middle of my talk, I looked up at the audience and I thought: I don't want to do this for the rest of my career.'

It was a decisive moment but at that time Jeremy had no idea what else he might do. Fortunately, however, a chance meeting with a very good friend, Dave Roberts, in the coffee room at the Weatherall Institute of Molecular Medicine in Oxford saved the day. Dave told him that the university was looking for a new Director for their Clinical Research Unit in the Hospital of Tropical Diseases in Ho Chi Minh City, Vietnam and, encouraged by his friend, Jeremy applied 'totally out of the blue'. He knew nothing about tropical diseases but he had always wanted to work overseas and loved the idea of returning to Asia.

He arrived in Ho Chi Minh City in 1995 when it was 'a town of bicycles and barely any street lights'. The Clinical Research Unit occupied a couple of small rooms at the back of the Vietnamese government's Hospital of Tropical Diseases and employed a handful of staff. As the name of the unit implies, medical research and clinical work went hand in hand. One informed the other. Understanding the way that pathogenic microorganisms behave makes it possible to develop better treatments and prevent the spread of infection or, at least, slow it down.

'At heart I'm a clinician,' Jeremy told Jim. 'As a doctor, your primary interest is what's happening in front of you every day.' Having done a PhD in basic science, he was also aware that to reduce suffering and minimise deaths, he needed to take a long-term view. 'I do strongly believe that one has to have a vision,' he told Jim. 'Blue-sky thinking about where the world is going is essential. Research needs to be purposeful. You need to imagine how you want the world to be twenty, thirty, forty years from now.' And this can be achieved, Jeremy believes, 'if you focus on the things of today'.

'Research needs to be purposeful. You need to imagine how you want the world to be twenty, thirty, forty years from now'

When Jeremy arrived in Vietnam, malaria was a major killer. 'Eight out of ten people [who had been infected by falciparum malaria] would have died with severe malaria and kidney failure'. When he left, eighteen years later, malaria was no longer a death sentence. 'I think – I hope, partially related to our work on understanding malaria, its treatment and critical care – the number of people dying in that condition would have been less than one out of ten.'

Patients in Ho Chi Minh City were treated with a traditional Chinese medicine, Qinhaosu, which led to medical advances that spread beyond Vietnam. Qinhaosu (which is now better known as Artemisinin) has since saved countless lives around the world. 'The research that we did on malaria, tuberculosis, HIV, Dengue, typhoid ... has global relevance. And I think it has changed – I hope it has changed – medical practice the world over.'

Medical professionals in Vietnam had more than enough work to do trying to keep all the *known* infectious diseases under control. And in 2003 a new threat to public health emerged: a novel coronavirus was causing Severe Acute Respiratory Syndrome (SARS). All the coronaviruses that were known back then were associated with nothing more serious than the common cold. This one was leaving many of its victims unable to breathe and killing one in ten of the people it infected. 'It was totally devastating,' Jeremy told Jim. 'I lost some good friends to SARS, doctors who looked after patients in Vietnam.' It came out of southern China to Hong Kong. 'Then it came to Vietnam and easily spread between people and particularly between patients and nurses and doctors.

'A very good friend of mine, Carlo Urbani, who was the World Health Organisation representative in Vietnam, made an incredibly brave decision, which should never be forgotten,' Jeremy said. 'He realised that something unusual was going on in the hospital he was based in – a hospital in Hanoi in northern Vietnam – and he advised the government to close it.'

Carlo's response to the arrival of SARS was not high-tech but it was effective. His plan was simply to stop people who were infected from infecting others. And he acted fast. He set up meetings with ministers and within weeks a task force had been created to deal with this crisis. Vietnamese government officials were instructed to identify and report any cases and anyone who was displaying symptoms was sent to one of two hospitals, the one Carlo had isolated and another designated SARS hospital. All of their contacts were tracked and traced and visited daily. It helped that SARS entered the country through one American businessman who had stayed at the Hotel Metropole in Hong Kong and that he was diagnosed promptly. Infra-red scanners were installed in airports and international travellers exiting the country had their temperature taken.

Many of the doctors and nurses who stayed inside the two quarantined hospitals became ill. Carlo lost his own life to SARS. But his prompt and courageous decision ensured that Vietnam averted what could have been a devastating epidemic of SARS. In total in Vietnam, there were sixty-three cases and six people died. An outbreak of SARS in Toronto killed twenty-four people.

'Remarkably, for a country at this stage of its development, and this GDP, Vietnam has an excellent national health service. If you go to a healthcare station in rural Vietnam you will find a nurse and you will find a doctor and you will find a community who uses that health system ... The very, very vertical structure of Vietnamese society and Vietnamese politics' might have helped, Jeremy thinks. 'I am not condoning the political system in Vietnam and the lack of plurality,' he said, 'but, in some ways, that vertical structure, with clear authority, has some advantages when you are trying to deliver top-down approaches.' With all the changes that have happened in the country since, whether 'it works for the future, is a different question', Jeremy said. But in 2003, clear lines of command and control 'from major hospitals in Hanoi and Ho Chi Minh right the way through to the community health stations in a village in the Highlands', resulted in health officials throughout the country delivering a prompt and co-ordinated response to this emerging public health crisis. Two months after the first case was identified in Hanoi, Vietnam became the first country to be declared SARS-free by the World Health Organisation: an impressive achievement for a country with such limited resources.

After the immediate threat had receded, medical professionals in Vietnam remained anxiously aware that this deadly virus could return at any time. 'We were all highly sensitised by SARS,' Jeremy said. The need to be extremely vigilant was uppermost in the mind of every health worker in the country. And so when in January 2004 a twelve-year-old girl living in southern Vietnam developed an unusual lung condition, the doctor responsible for her care immediately alerted the local World Health Organisation official. Without missing a beat, the new WHO representative in Vietnam, Philippe Calain contacted the Hospital of Tropical Diseases to alert them about this suspected case of SARS.

'I remember it well,' Jeremy said. 'It was the Vietnamese New Year, Tết – imagine Christmas, Easter and

Thanksgiving all rolled into one ... Everyone was on holiday. The hospital was practically deserted.'

Jeremy and his colleague, Professor Tran Tinh Hien were on duty. Hien and the WHO representative went to visit the girl in hospital straightaway and conducted a medical examination. Once the WHO doctor was satisfied that this was not a case of SARS, he left. Hien had a sixth sense something was odd. He stayed on and took a detailed case history. 'Hien was really key here,' Jeremy said. 'He is a brilliant clinician.' When the girl told Hien how her brother had buried a duck and she had dug it up, alarm bells rang. Hien knew ducks had been dying from a particularly virulent form of bird flu, H5N1, and that people who came into close contact with infected birds were at risk.

This avian influenza had infected eighteen workers, and killed six, in a live poultry market in Hong Kong in 1997. The Hong Kong government had acted swiftly and successfully contained this outbreak by killing all the chickens in the territory. But there had been another outbreak in Fujian province in China in 2003. It didn't get much media attention because SARS was the story at the time, but Hien had read about it in the academic literature and was aware of the serious threat that H5N1 posed to human health.

He came back to the hospital on his motorbike with a blood sample in his rucksack and alerted Jeremy, late on the Eve of Tét. Hien agreed with the WHO doctor that it wasn't SARS. But wondered if the girl's breathing problems were caused by H5N1? If this was the case, he knew he would need to act fast. With a highly pathogenic virus like H5N1 you don't leave anything to chance. Jeremy's wife, Christiane Dolecek, who was also a doctor at the

Hospital of Tropical Diseases, left the family Tết party and came to the hospital to run the molecular tests for a range of infections including H5N1. And so it was that while the rest of the country was cele-

'With a highly pathogenic virus like H5N1 you don't leave anything to chance'

brating the New Year, the first human case of H5N1 avian influenza was diagnosed in Vietnam.

'That is what set the whole thing off,' Jeremy said. Just as the county was still reeling from the shock of SARS, another virus that was new to Vietnam, and highly pathogenic, threatened to take yet more lives. H5N1.

'It was ten years ago and I can still remember it minute by minute ... Some things just remain etched on your mind, as if it were yesterday. SARS had hit the north of the country, a good friend Carlo and many others had died. This time the Hospital of Tropical Diseases in the south was the epicentre and Hien, Christiane and Jeremy were in the hot seat.

'Seeing the number of cases escalate daily' was scary and a terrible reminder of all the people who had died from SARS, in particular his great friend, Carlo Urbani. There was 'clearly something very serious going on' and it shook him to the core. 'It does make you question your profession. Your morals. Your attitude to risk,' he told Jim. As a doctor, he was programmed to treat the patients and to look after the people in his care. As a researcher, he knew it was critical to ask the right questions and find solutions that were 'important for Ho Chi Minh and – potentially – the world'. At the same time, he had a family to look after and protect. 'Working out how to put all of those things together' was difficult. 'It was a very emotional time. Still is, in many ways. I did not go home for two weeks. It was very frightening,' Jeremy said.

After the girl was diagnosed, twenty-nine more cases were identified in Vietnam, twenty of whom died. H5N1 has, on average, a 70 per cent mortality rate but the girl made a complete recovery. She visited the hospital five years later just as she was about to go to university. There was another outbreak of H5N1 in Vietnam in 2005. This time more people were infected but fewer of them died, lessons had been learnt in the hospital on how to treat these patients and it made a difference. H5N1 came back again in 2007, 2008, 2009 and 2010 and remains a risk to human and animal health to this day. The first case of human-to-human transmission was identified in 2006. Mercifully, to date, transmission between people has happened only rarely, but virologists are watching carefully in case this changes as the virus mutates.

> 'It does make you question your profession. Your morals. Your attitude to risk . . . It was a very emotional time. Still is, in many ways'

The secret of Vietnam's success in containing outbreaks of H5N1 and SARS was to be on the front foot, with a clear coherent and consistent strategy. The speed with which health workers in Vietnam responded to cases being diagnosed made a real difference to what happened next. If transmission can be stopped early on and outbreaks contained, there is less need for state-of-the-art medical care and hospital beds. Saving lives doesn't have to cost a fortune.

To win the war on infectious diseases, we need to be well informed. And we need to be alert.

And, Jeremy believes, we also need to start to think about how *we* live.

The way we interact with the changing environment, and ecology in different parts of the world, all directly impact on human and animal health. 'Urbanisation, people moving to the cities; how they use land, their interactions with animals – whether it be invading forests to cut down wood, and becoming exposed to animals one would not normally come in contact with; or the way we farm – all these things will have profound implications for the emergence of new infections.'

Understanding societal changes like these requires us to have insight into areas that would not normally be considered part of a programme to tackle infectious diseases. It's too easy to blame 'a deadly virus' for an outbreak of disease, without asking what we might have done to allow these microorganisms to thrive. Our actions have a significant impact on the survival chances and behaviour of *every* organism on the planet – microorganisms that are pathogenic to human health included. To protect our health, we need to protect the health of our animals. We need to think about the big picture and to see how everything fits together. We can't take ourselves out of the equation. Because *we* might be part of the problem.

In October 2013, after eighteen years in Vietnam, Jeremy moved to London and became the director of the Wellcome Trust, a charitable foundation that is committed to improving global human health through research. With an annual budget for research of £1 billion, the Trust is one of the largest independent funders of scientific research in the world. Medicine tends to focus on the problems that are presented by patients. But Jeremy is adamant that solving our health problems demands that we maintain the broadest possible focus. 'The humanities . . . are critical, critical bits of what we need to support,' Jeremy said. 'We have had a long-term commitment to working in the humanities, in culture and society. I think the Trust has been way ahead of the game [in that respect].'

Working as 'a very junior doctor' at St Stephen's hospital in London (now Chelsea and Westminster Hospital) in the 1980s, Jeremy saw 'young people, mostly males at that stage, dying of [what was then] an untreatable infection, HIV AIDS . . . That experience has stuck with me for the rest of my life,' he said. Serendipity took him to the Hospital for Tropical Diseases in Vietnam in 1995. In those days, the Clinical Research Unit in Ho Chi Minh City employed a handful of people. Under Jeremy's leadership it grew into a thriving state-of-the-art research facility employing hundreds of scientists in Vietnam and with outposts in Indonesia and Nepal.

'Many people over the last fifty years have said infectious diseases are going to be something of the past, vaccines and antibiotics will push them to the side.' They couldn't have been more wrong. HIV reminded us that 'new things can emerge'. And that science has the power to transform

> 'Many people over the last fifty years have said infectious diseases are going to be something of the past, vaccines and antibiotics will push them to the side'

lives. 'Here was a new infection for which there was no treatment and no vaccine ... but within a few years we had treatments which turned a life sentence of a few weeks after diagnosis, to a chronic condition that could be managed. HIV was no longer a death sentence. People living with HIV could lead a full life.'

Ebola, Dengue, Lassa, Nipah, Zika, bird flu, Swine Flu, MERS, SARS and Covid-19 are all threats to human health. With a changing environment, urbanisation and ever more international travel, these infections when they emerge or re-emerge will not respect borders, what happens in Asia today will impact Europe and America tomorrow and vice versa. And then there's the looming spectre of rising antimicrobial resistance. 'Drug resistance shows you that old diseases will come back ... We will see that coming back to haunt us. Everybody who dismissed infectious diseases has looked back on what they said with regret,' Jeremy told Jim five years before the arrival of Covid-19. 'So infectious disease will remain a huge part of the work that Wellcome supports.'

> 'Everybody who dismissed infectious diseases has looked back on what they said with regret'

Thank goodness for that.

WENDY BARCLAY

Fighting flu

Born: 1964

Grew up in: South East London

Occupation: virologist

Job title: Professor of Virology at Imperial College London

Viruses studied: rhinoviruses, polio, influenza viruses, SARS CoV-2

Inspiration: 'seeing an astonishingly detailed and beautiful image of Rhinovirus 14'

Passion: 'taking viruses apart to see how they work'

Mission: 'to use molecular virology to protect us all from the viruses that cause us harm'

Advice to young scientists: 'be open-minded'

PROFESSOR WENDY BARCLAY studies the molecular machinery of viruses and uses this understanding to assess why some viruses are more deadly to humans than others. Her research career began analysing nasal swabs taken from snotty human volunteers who were enjoying an all-expenses-paid holiday with a one-in-three chance of catching a cold, at the Common Cold Unit near Salisbury. She then specialised in influenza viruses, and worked out how to make them in the laboratory so that they could be studied in more detail. She has studied the Influenza B viruses that are responsible for seasonal flu and Influenza A viruses found in animals that have the potential to cause pandemics. H5N1 bird flu, for example.

Wendy talked to Jim Al-Khalili in January 2018 during a particularly bad winter flu season.

The flu is a formidable foe. Influenza viruses are tiny. They can't even reproduce without a host. And yet they are responsible for hundreds of thousands of deaths every year. How do these simple microorganisms manage to destroy large, sophisticated and intelligent multi-cellular creatures like us?

Wendy Barclay has spent most of her life trying to answer this question. 'It's this David and Goliath battle that the virus often wins,' she told Jim. She studies influenza viruses, seeking to understand how they operate, right down

> 'It's this David and Goliath battle that the virus often wins'

at the level of individual molecules. She didn't enjoy the study of organisms at school, however. The concepts in biology were 'big and difficult to probe', whereas the physical sciences enabled her to break the world down into 'small, understandable chunks'. She studied Natural Sciences at Cambridge University and graduated with a degree in pharmacology, having been encouraged to 'take some modules outside her main area'. And applied for all sorts of marketing jobs. 'I don't know why I thought that's what I should do,' she told Jim. 'I was completed unsuited to sales!' When the marketing departments of all the major pharmaceutical companies turned her down, she pursued plan B: a job (and a PhD) studying rhinoviruses at 'a place she'd never heard of', the Common Cold Unit near Salisbury.

'I realised, after I accepted the offer, that I didn't really know what a virus was!' she told Jim, smiling. Fortunately, these tiny microorganisms were right up Wendy's street. They satisfied her deep-seated desire to be able to get to the bottom of things.

'I remember the late David Tyrrell sent me a paper that had just been published in *Nature*,' Wendy said. The paper contained some stunning images of Rhinovirus 14, one of the viruses that causes the common cold. It was the first time a virus had been described in such astonishing detail. 'And I looked at it and could see that this was the sort of science I liked,' Wendy said. 'It was very small and you could understand it all. You could even see it. You could see where every single molecule in the virus was. And I thought yes, I can get to grips with this!'

'You could see where every single molecule in the virus was. And I thought yes, I can get to grips with this'

The Common Cold Unit was a medical research centre with a difference. 'It was basically a network of huts connected by wooden runways' which provided temporary homes for human volunteers and people came from all over the country, having been promised 'a cheap and comfortable, all-expenses-paid holiday in the Wiltshire countryside'. Three meals a day included. Many volunteers came back year after year. For some, a visit to the Common Cold Unit was a way to honeymoon on the cheap. There was a one-in-three chance of catching a cold but visitors were reassured that infections were 'in a very good cause, and usually minor and brief'. And, over the years, 20,000 people enjoyed a free, ten-day autumn or winter break that, they felt, was not to be sneezed at. In return, the scientists who worked at the unit gained access to a lot of snotty humans and a steady supply of the viruses that cause the common cold. Dozens of rhinoviruses were discovered at the CCU and, in 1965, the Director David Tyrrell and visual-imaging pioneer June Almeida described the first coronavirus, so-named because it looked as if it was surrounded by a crown. But studying the viruses that cause the common cold was considered a luxury that the funders of medical research could not afford when the HIV AIDS epidemic hit and the unit was closed down in 1990. With the benefit of hindsight, it seems learning more about the coronaviruses that cause common colds might have been quite useful. Even if a vaccine for the common cold remains elusive.

'The volunteers would be put into quarantine for a few days to check that they hadn't brought anything with them, a cold or flu that we didn't know about,' Wendy explained to Jim. And provided they didn't develop any symptoms over the first two or three days they would then be given a common cold virus or a placebo. Wendy's job, and the subject of her PhD, was to investigate whether catching a cold one year stopped you from catching cold again. Or, more precisely, did being infected with a particular cold virus one year confer immunity against that virus the following year? She worked with the volunteers and spent a lot of time taking nasal swabs. It wasn't the most pleasant aspect of her work, but was preferable perhaps to some of the methods that had been used in the past which included picking up tissues that had been used by the volunteers to blow their noses. 'My volunteers came and I gave them all a known rhinovirus. The following year I persuaded them to come back again and I measured the antibodies in their blood and in their nose to see if they were still, in theory, immune.' If antibodies were present, she would then give those volunteers the same virus again to see if the presence of these antibodies stopped them from catching cold again. In theory they *should* be immune but every scientific theory needs to be tested.

Having completed her PhD on rhinovirus immunity, Wendy got a job doing research with Jeff Almond at Reading University. Jeff was a leading figure in an exciting new field of molecular virology that had emerged in the US in the early 1980s. 'He was using what were then pioneering techniques to sequence the genomes of small viruses like the rhinovirus and other related viruses,' Wendy said. Viral genomes are short and simple, which makes them a lot easier to study than the human genome, for example. (Flu viruses typically have about twelve genes. Humans have more than a hundred thousand.)

'Not only that,' Wendy said, 'he was able to use a technique to generate viruses from scratch in the lab.' He followed the code (that he had revealed by sequencing the genome) and used it to make a synthetic version of the genetic material of the virus, known as RNA. These viruses don't have DNA. Instead, they hijack the machinery of the cells that they infect and use their RNA as a template to make copies of themselves. Jeff then put this piece of reconstructed RNA into human cells and, as if by magic, it behaved as if it was the real thing. 'Boom! The next day the cells were full of thousands and thousands of viruses, all of which are derived from that one piece of laboratory-manufactured replicating RNA. This was amazing to me!'

Wendy said. 'I was absolutely blown away.'

It was a virus-making factory. Jeff had found a way to create viruses from synthetic RNA, so new viruses could be custom-made to order. And this ability to genetically engineer viruses revolutionised our ability to understand how they function. 'You could redesign the virus,' Wendy said, excited. 'And then you could see how these genetic changes altered the behaviour of the virus.'

> 'You could redesign the virus. And then you could see how these genetic changes altered the behaviour of the virus'

Flu genes contain the instructions for making different proteins – these are the molecules that enable a virus to do what it does. Surface proteins, for example, are the means by which a virus attaches itself to the host's cells and gets inside them. Other proteins are responsible for hijacking the host machinery and persuading the host to make more copies of the virus rather than its own cells. 'So you could conduct hypothesis-driven science where you look at one piece of the genome and say, I wonder what that's doing?' Wendy said.

While Wendy was manufacturing the RNA of the rhinoviruses and poliovirus, Peter Palese at the Mount Sinai Medical School in New York had started trying to make the viruses that cause flu. And when he heard Wendy give a talk at a conference, he offered her a job. Flu viruses are more complex than rhinoviruses, their genomes are made up of eight physically separate pieces that must all be delivered into the same cell. Not only that, but the RNA is carried into the cell in what is known as the negative sense. This means it is not infectious in its own right, but needs other flu proteins to help it. Nonetheless, Peter Palese had overcome these hurdles and managed to manufacture Influenza A viruses. Wendy joined his laboratory in New York and was the first person in the world to create synthetic Influenza B viruses. These are the ones that are responsible for seasonal flu.

'The question that taxes me at the moment,' Wendy said, 'is why are some viruses so deadly, and others not?' In particular: 'Why does one strain of flu behave differently from another?' And the key to answering this question is understanding the genetic differences between them. Flu viruses are constantly evolving – and at an alarming rate. Significant changes can occur in a virus population in a matter of weeks and it's this ability to change so fast that makes influenza viruses so difficult to outwit.

As soon as a vaccine is developed to counter one strain, another one has emerged, which may be more virulent than the last. The annual flu jab can become out of date in the time it takes to be manufactured, typically about six months. 'If we use drugs against it, it can wriggle out of that as well.' Flu viruses mutate their way out of trouble and render redundant pharmaceuticals that were cleverly designed to disable their former selves.

'Given that it's so hard to predict which virus is going to make people ill, how on earth does the World Health Organisation decide which strain to grow the vaccine against?' Jim asked.

'It's a really difficult task,' Wendy said. 'Because all the flu viruses all over the world are all out there, mutating . . . It's a moving target, so you just hope that you got it right.'

'We have a huge network of surveillance laboratories

that pick up viruses from all around the world and sequence their genes.' They concentrate on the genes that encode surface proteins, the ones that enable the virus to enter the host cells, to see if there is any change. All this information is then 'brought

'. . . all the flu viruses all over the world are all out there, mutating . . . It's a moving target, so you just hope that you got it right'

to the table at the WHO in about February. Everybody sits down and makes their – I won't say it's their best guess – makes their informed decision about which of the strains are most likely to be causing flu the following winter. The vaccine manufacturers then have about six months to make that product, manufacture it, put it in bottles and get it out to GPs. Because we've got to get it into people's arms in order to induce antibodies before the next flu season arrives.

'And all the while, between that decision being made in February and the flu virus actually coming back the following December, the flu viruses are carrying on mutating' and new strains keep emerging. Even if the WHO got it right in February, 'the virus may have run away from us by the time it reaches us in December' when the new flu season arrives.

With so many viruses all mutating, it is hard for us humans to keep up with 'these little packets of genetic information that self-perpetuate and move around'. Working out what flu will do next is, 'to paraphrase Winston Churchill, a riddle wrapped in a mystery, inside an enigma', Wendy said. Winston was referring to the actions of Russia at the beginning of the Second World War but she feels the same about the flu. 'Influenza viruses are the classic example of the rapid evolution of a virus. And that's a major challenge.' Rapidly mutating viruses put the people who are trying to make vaccines on the back foot and always playing catch up. In an ideal world we would be able to cover all bases, up front.

'Will there ever be a universal [influenza] vaccine?' Jim asked. 'That's the dream of many influenza virologists,' Wendy said. 'Often the media like to pick up on the research that's working towards it' but it is not an easy thing to achieve. 'There are a couple of different versions of the universal flu vaccine and they all work in ways which are very different from the current vaccines that we use.' At best these vaccines will be able to limit the infection and there is a chance that the vaccine itself could make us ill. (The annual flu jab is safer because it stops flu viruses from entering our cells in the first place.) Then there's the 'huge technical challenge' of producing a new vaccine product. 'And you've also got to think about how we would give those vaccines and who we would give them to.'

The annual flu jab was originally recommended only for the elderly and those with certain underlying health conditions, because they were the most at risk of serious illness or death. 'One of the things we do differently now,' Wendy said, 'is we vaccinate children. There is a new vaccine in the UK, the live-attenuated vaccine, which we spray up children's noses. This was an idea that had been around for a while but an analysis of the data collect-ed during the Swine Flu pandemic in 2009 prompted a change of policy in the UK. The number of people who were going to their doctors with suspected or even con-firmed Swine Flu was rising from around May through June and into July. Then, in the third week of July, the numbers just went away. And, of course, what happens in the third week of July is that the school children break up for the summer holidays.'

This evidence suggested that 'Swine Flu spread through the community via [asymptomatic] schoolchildren who mingle together.' And soon after these results were published, a new vaccine to protect the youngest members of society from seasonal flu was introduced in the UK. Most children don't suffer too much from the flu viruses that circulate every winter, but giving them immunity helps to protect us all.

Humans have lived with seasonal flu viruses for centuries. We have built up immunity and have vaccines that offer us some protection against them. But the viruses that circulate every winter (which are mainly Influenza B viruses) represent a fraction of all the influenza viruses on earth. The Influenza A viruses that infect animals (mainly birds) represent another threat to humans' health, and since these viruses are totally new to humans, our immune systems have not been trained to spot them. These are the influenza viruses that have the potential to cause pandemics.

Avian influenza viruses usually have a hard time infecting humans. They have not coevolved with us as their hosts and so their molecular machinery generally does not operate very well in human cells. But they too are capable of rapidly evolving and, as they mutate, they can develop new ways to enter the cells of different hosts including humans. A further mutation might allow them to take over the human cell machinery for their own replication. The final challenge for these animal influenza viruses is to find a way to move between human hosts. This could be achieved by travelling in our bodily fluids – snot, or saliva – or by finding ways to move through the air we breathe. If all these three challenges are overcome by the virus, then we have a problem.

The Spanish Flu pandemic (1918–20), the Avian Influenza pandemic in 1957 and the Hong Kong Bird Flu pandemic of 1968 were all caused by bird flu viruses jumping across into humans. Even the Swine Flu pandemic (2009/10) had its distant origins in avian viruses but was mercifully mild by comparison. 'Most people in London had had Swine Flu by September without even knowing that they were infected,' Wendy said.

Influenza A viruses are classified according to the H and N proteins that are found on the surface of the viruses: H_1N_1, H_2N_2, H_3N_2 etc. The H and N proteins are what give the virus its distinctive shape and unique ability to attach to and invade human and animal cells. To date, only viruses with certain combinations of H and N surface proteins have caused human epidemics. And, courtesy of some clever retrospective genetic detective work, we now know which ones they are.

A virus found in samples of lung tissue that had been taken from the victims of Spanish Flu in 1919, and stored in a freezer ever since, has been analysed using modern techniques and the sequence of the RNA suggests it was an H_1N_1 virus. A century later, a modified version of the H_1N_1 virus found its way into humans once again, this time via pigs, causing the Swine Flu pandemic in 2009.[1]

---

1 Unusually for a flu virus, most of the people who became seriously ill with Swine Flu were young. It's thought the elderly had developed or inherited some immunity from Spanish Flu, because they had been infected in early life by a version of Spanish Flu that shares the same H surface protein.

'What's really fascinating,' Wendy said, 'is that the genes that make up that virus appear to have come from two different places. Two different viruses that were living in pigs who were living perhaps on two different continents. Quite how those two pigs got together, I think we'll never know. We know that the first few cases were in the southern US, and also in Mexico. We can imagine a scenario where, due to farming practices, pigs are being moved around the world and some pigs carrying two different viruses had been brought together. The viruses might have co-infected a single pig and mixed their genes up in a special way that flu is rather good at doing. And this shuffled-up virus comes up now with the ability to jump into humans and transmit through the air between people. And off we go, we've got a massive pandemic.'

'It doesn't sound like this was something that we could have been prepared for,' Jim said.

> '. . . shuffled-up virus comes up now with the ability to jump into humans and transmit through the air between people. And off we go, we've got a massive pandemic'

'That's right. And that's why it's so difficult,' said Wendy.

Following an outbreak of bird flu in 1997 which killed six of the eighteen people it infected at a live poultry market in Hong Kong, the genome of the virus responsible was analysed and its animal origins traced. The genetic evidence suggested that their death was caused by a highly pathogenic strain of H_5N_1 found mainly in Asian wild birds. This wild bird virus infected chickens which in turn had infected humans who had come into close contact with them. The Hong Kong authorities slaughtered a million chickens as a precaution and this outbreak was contained. But six years later there was another outbreak of H5N1, this time on mainland China, and many were concerned.

Exotic birds in a park in Hong Kong and wild birds such as geese were dying. Scientists knew that a lethal strain of H5N1 was being spread across continents by migrating wild birds and the commercial movement of domestic poultry, and almost half a century after the avian influenza pandemics of 1958 (H2N2) and 1968 (H3N2), many feared another avian influenza pandemic.

'We were expecting a bad outbreak of bird flu back in 2004/5, when the circulating strain was H5N1,' Jim said. 'How did we stop that getting a hold?'

'I don't think we can take too much credit for that,' Wendy said. 'The H5N1 virus is still out there in South East Asia, and is actually widespread across nearly all the continents of the world, in Africa, and in parts of Europe. But I think we should give credit to some influenza researchers who rang the alarm bell back in the early 2000s about viruses like H5N1.'

The WHO pandemic advisory committee made medical professionals around the world aware of the great threat that this apparently lethal strain of bird flu virus posed to humans. If this virus were to find a way to move between humans, there would be terrible consequences.

'The reasons why H5N1 is unable to transmit efficiently between humans are a matter of intense research,' Wendy said. 'Many possible explanations have been suggested and researched, including studies of animal and human behaviour and ecological changes. It's hard to come up with a definitive answer, when there are so many factors at work,' she said. But Wendy thinks that there are 'real barriers that make it very difficult for the virus to jump into humans'. H5N1 has not found a way to achieve airborne transmission between humans, at least not yet. And detailed studies of how this virus works have shown that the odds are stacked against it mutating in this way. 'It would be like rolling three dice and each one of them being a six,' Wendy said. Given what we now know about the strengths and weaknesses of the H5N1 virus, it seems highly unlikely that H5N1 could cause a pandemic.

'Why was Spanish Flu so deadly?' Wendy said. 'Or the bird flu viruses which sometimes jump across from poultry into humans in parts of Asia? Why are they so lethal? Why are some of these viruses so deadly and others not? What is it about the gene sequence that makes that difference? These are the kinds of questions we are still trying to address.'

'And now there's a new one, called H7N9, in China.' When this virus was first detected in humans in 2013, Wendy and her team set to work. 'I think we've come on leaps and bounds in understanding that virus . . . Some of the research that we've done in my group is to identify the differences in the virus's proteins between strains of the virus that are found in humans and in birds. We can now explain why bird viruses don't replicate well in human cells, unless they undergo certain mutations.'

Other groups are working on why some viruses are able to travel through air. 'We understand much better why some bird flu viruses struggle to survive in the air

> 'We understand much better why some bird flu viruses struggle to survive in the air and then penetrate into the next host, even if they do land in an infectious form in somebody's nose'

and then penetrate into the next host, even if they do land in an infectious form in somebody's nose. We really do understand a huge amount more than we did ten years ago,' she told Jim. 'Of course, that's always the way with science. But whether or not that's going to enable us to predict the next influenza pandemic, I think, *that* is the question . . . Because, to be honest, we can live with the flu. We can manage it, if it's a mild version. But if we have an Influenza A virus jumping across, in that deadly manner, then that's going to be a different game altogether.

'Knowing which of the Influenza A viruses living in animals, mainly in birds, are going to make the jump across to humans . . . is a bit like looking for a needle in a haystack,' Wendy admitted. But it hasn't stopped her, and many others from trying. 'Understanding that jumping process is one of the major challenges in influenza research,' Wendy told Jim in 2018. For a long time, bird flu viruses were thought to represent the biggest threat to human health. An Influenza A virus was expected to be the cause of the next pandemic, not a coronavirus. Nevertheless, the skills that Wendy developed to study viruses that cause respiratory illnesses put her in a really strong position to study SARS-CoV-2 when it was identified as the cause of Covid-19.

JONATHAN BALL

*Developing antibody-based
treatments for HIV and Hepatitis C*

Born: 1965

Grew up in: Clipstone, a mining village in Nottinghamshire

Occupation: virologist

Job title: Professor of Molecular Virology at the University of Nottingham

Main viruses studied: Ebola, HIV, Hepatitis C

Inspiration: 'growing influenza on hens' eggs'

Passion: 'genetically engineering viruses and antibodies'

Mission: 'to tackle viruses head on with treatments and vaccines'

Advice to young scientists: 'do what you enjoy, not what's expected of you'

PROFESSOR JONATHAN BALL is interested in how viruses evolve and has worked on antibody-based treatments for HIV and Hepatitis C. Early in his career, it occurred to him that there might be different strains of HIV living in different parts of our bodies and he was one of the first people to study HIV in the genital tract. Viruses like HIV and Hep C mutate at an alarming rate and finding ways to disable them is a real challenge. The approach Jonathan adopted to defeat Hep C was to target those bits of the virus that didn't change. During the West African Ebola epidemic (2014–16) he studied how the virus changed as it infected more people. Did it become more deadly?

Jonathan talked to Jim Al-Khalil in July 2019.

At the Centre for Applied Microbiology & Research laboratory in Porton Down, 'if you hadn't turned up by half past nine, the police would be sent to bang on your door to make sure that you weren't lying dead somewhere', Jonathan Ball told Jim. 'It was a place that dealt with a real serious threat' from deadly microorganisms such as anthrax, Lassa and Ebola and where entire corridors would be shut off and fumigated by a team of cleaners dressed in full bodysuits, masks, gloves and boots.

But working in a high-security laboratory didn't feel dangerous to Jonathan. He enjoyed being part of a community where everyone knew everyone and 'researchers worked together for the common good'. It reminded him of 'the pit environment' in the mining village in Nottinghamshire where he grew up. 'The pit was the whole focus for the community,' he said. 'Most of the families had some association with the mine. If they weren't deep underground, they were working on top ... everyone looked out for one another ... you just felt safe in that community.'

Aged sixteen, one option open to Jonathan was to follow his brother and his dad, down the pit. The view from upstairs in the family home was of 'four or five different pits ... that's what the expectation was', and he'd seen all the banknotes that his older brother 'stuffed into the kitchen drawer'. But over the summer he decided to stay on at school, safe in the knowledge that a job in the pit would always be there. 'It was maybe not the most academic of environments. And I don't mean that in a derogatory way' he said. 'There were certainly some very, very bright people, but I don't think the expectations were there for us to go on and get A levels . . . The sixth form was very small. School was there pretty much to keep us out of mischief until we were sixteen and old enough to go to the pit. Or – if you were a girl – you'd probably end up going to the hosiery mills or the factories or shops.

'I was the only one studying Chemistry A level, and I'd often teach myself out of a book at the back of the O level class.' And he managed to earn quite a lot of money playing the organ at funerals. 'If you could learn "Abide With Me" and "The Lord Is My Shepherd", you were pretty much a gigging funeral player.' When funerals clashed with chemistry classes and he played truant from school, he had the perfect excuse. 'Where have you been?' his teacher would ask. 'At a funeral,' Jonathan would reply, feeling very smug.

His parents were happy for him to do whatever he wanted as long as he stayed out of trouble. But a couple of teachers urged Jonathan to think about what he was going to do with his life and, suitably nudged, he went off to study Applied Biological Sciences at Bristol Polytechnic. One of his lecturers, Professor David Shaw, alerted him to the fact that studying viruses could change the world and an undergraduate project growing influenza viruses on hens' eggs got him started in the field.

}

When the HIV AIDS epidemic arrived in the UK, Jonathan was working as a research assistant at the government laboratory in Porton Down. At that time there was no treatment for HIV AIDS, no cure. Leaflets were dropped on every doorstep in the country in January 1987 with a terrifying picture of a tombstone and a warning 'AIDS: don't die of ignorance'. The message was clear: to have sex without a condom was to risk your life.

The Centre for Applied Microbiology and Research was one of a very few places in the UK where such a hazardous virus could be safely studied: the scientists at Porton Down had been studying 'some of the nastiest viruses and bacteria for years'.

'We were interested in developing a vaccine, but before you can do that, you have to understand the basics of the virus. And one of the things that this involved in the early days was to generate sufficient virus for us to be able to visualise it.' Jonathan did this

> 'We were interested in developing a vaccine, but before you can do that, you have to understand the basics of the virus'

by 'infecting lots of different flasks full of human cells' to produce 'lots of culture fluid'. Then the challenge was 'to somehow concentrate the virus from that fluid'. They did this by suspending the liquid in a sucrose solution and spinning the samples at high speed in a centrifuge. This separated out the contents of the fluid according to their relative densities and 'eventually the virus particles formed a band. We could see a pure layer of virus . . . There were literally millions and millions of virus particles in that layer,' Jonathan said, still sounding excited decades after the event. Seeing a coloured band of liquid that he knew

> 'I knew for a fact that, if I made a mistake, there was a chance that I could become infected with HIV'

was teeming with HIV, the invisible killer had been made visible. 'And I remember shaking it, harvesting the virus. I knew for a fact that, if I made a mistake, there was a chance that I could become infected with HIV. And that really made me think: "This is something important." Because if it could kill me, it could kill anyone out there.'

Once the virus had been isolated, the next step was to try and understand how it behaved. Jonathan's boss, Peter Greenaway, did this bit. 'He was the one who inspired me,' Jonathan said. 'He was a genuinely nice guy and one of the first scientists in the UK to work on recombinant DNA.' Thirty years before the invention of the sophisticated gene-editing techniques (like CRISPR) that are increasingly used today, he found a way to manipulate the genome of viruses like HIV, by first cloning the virus, then breaking up its genetic material and recombining different bits. This was where the action was, Jonathan thought and he wanted to be part of it. 'I wanted to get some training and experience in what I thought was going to be the future – this genetic engineering … If you can manipulate a genome you can start to understand how a virus behaves.'

❧

The experiments that were carried out at Porton Down involved viruses that had been manufactured on location and studied in isolation. Jonathan's next job gave him the opportunity to 'really understand what's happening in a human who is infected with a virus'. He was working and studying part time for a PhD at Birmingham Heartlands Hospital, where the patients were 'always very willing and very keen to be part of these sorts of studies'. And so he had access to plenty of blood samples taken from people who had been infected with HIV.

His research was going well until, in the final year of his PhD, Dr Ulrich Desselberger, who had hired Jonathan to work in the hospital in Birmingham, moved to Cambridge. 'Leaving you with little choice but to follow him or go it alone and secure your own funding,' Jim said. 'You took the latter option and decided to go it alone. That's not just a brave thing to do. It's almost unheard of!'

'Naive, I would say!' Jonathan exclaimed. 'I didn't know any different. So, I put together a couple of grant applications and sent them to the Medical Research Council.' A bold move for a student who had not yet finished his PhD.

'I was summoned to go and visit a very esteemed colleague. He had my grant application in front of him and he almost threw it at me on the desk. Then he said, "Why should we fund you?" Basically, who do you think you are! I was quite taken aback by that,' Jonathan said. Although in hindsight he could 'see what he was getting at': PhD students don't normally submit grant proposals. But Jonathan succeeded, in part he thinks because the funding climate was favourable. HIV posed such a terrible threat to society that all proposals for AIDS-related research that scored sufficiently highly were automatically approved. And also, he admitted when pushed, 'there was the germ of a good idea in there as well'. He succeeded in gaining the funding he needed to set up his own research project on HIV AIDS and moved to Nottingham University, to work not far from where he had grown up and be near one of the leading lights in HIV research, Paul Sharp.[1]

Every time an HIV particle replicates, its genetic material is copied. It's not a perfect process and sometimes – quite often, in fact – mistakes are made. Other organisms have proofreading mechanisms in place to reduce the risk of copying mistakes. But for viruses like HIV, these copying mistakes are the reason it is able to evolve so fast. Most of the mutant viruses that are generated will perish, but if a mutant virus is created that is better adapted to living in a particular part of the human body then this well-adapted mutant will survive and thrive.

Jonathan wondered if 'different parts of the body could harbour different strains of the virus', just as the finches that Charles Darwin observed in the Galapagos Islands adapted and evolved in different ways to fill diverse eco-logical niches. Over time some finches evolved to have long thin beaks that were perfect for snagging insects, while others gained strong wide beaks that were excel-lent for cracking nuts. Would HIV adapt and evolve in different ways depending on the human tissues in which it found itself? Lymph, liver or blood, for example. And if so, might this 'influence how AIDS came about'? Most HIV researchers at the time focussed on viruses that were found in blood. It made sense. Blood tests and sampling are commonplace in hospitals. It's one of the easiest ways to track the progression of a disease. But as medical un-derstanding of HIV progressed, it became clear that the

---

1 Together with colleagues in the US, Paul worked out that HIV most probably jumped from chimpanzees to humans.

lymph system played an important role in the development of HIV AIDS. Jonathan 'wanted to find out how HIV was changing and mutating when it infected the lymph tissue, or the brain'. And he thought it could be useful to study the viruses that lived in the patient's genital tract. Since HIV AIDS is primarily a sexually transmitted infection, these were the virus particles that needed to be stopped in their tracks.

'I wondered if different parts of the body could harbour different strains of the virus'

Jonathan showed that the strains of HIV that were found in patients' genital tracts were indeed different from the strains that were found in their blood. He studied how these viruses worked and hoped to develop a vaccine that could stop them in their tracks and prevent the virus being passed from one person to another when they had sex. Genital tract HIV remains an active area of research but there are no easy answers. Different strains of HIV are found in different individuals and a constant process of reinvention means these targets keep changing. The more we study HIV the more complicated it becomes and the possibility of developing a vaccine for HIV AIDS now feels more, not less, distant than it once appeared.

In the mid-1990s, however, the treatment of HIV AIDS was transformed by highly active antiretroviral treatment (HAART), a drug therapy that allowed people with the virus to look forward to a

'I wanted to find out how HIV was changing and mutating when it infected the lymph tissue, or the brain'

near-normal life-expectancy. And when being HIV positive was no longer a death sentence, Jonathan refocussed
his research onto another blood-borne virus, 'the poor
relation, in a way, to HIV' – Hepatitis C. The main transmission route for this virus was through sharing needles.
Thousands of people in Nottinghamshire were injecting
drugs daily and by 2003 one baby a month was being
born already addicted to heroin.

The death toll from Hep C was huge. Hep C infected
200 million people worldwide, and yet had remained
'fairly unexplored.' Jonathan thought maybe he could
apply some of the many
lessons that had been learnt
about HIV to this neglected
area of research. 'I assumed,
naively in retrospect, that Hep
C would behave in a way that
was similar to HIV.' Guided by
this assumption, he set up a research programme to study
Hep C, 'using the same techniques, tools and approach
that had been used to study HIV'.

'I assumed, naively in retrospect, that Hep C would behave in a way that was similar to HIV'

Dr Will Irving, a Consultant Clinical Virologist had
assembled one of the largest cohorts of patients with
Hep C and Jonathan's plan was to try and understand
the immune response in people who were infected with
this virus. What antibodies did they produce? Hep C is
another virus that mutates rapidly and as with HIV this
was not an easy problem to solve, as the answer would
keep changing. In short, there was an arms race between
Hep C and the patient's immune system. Every time Hep
C made an advance, the patient's liver would incur more
damage, leading to a loss of liver cells and scarring, which
in turn prevents the liver from working properly, and

ultimately liver failure, which can be fatal.

When faced with an invading microbe our bodies normally mount a defensive response designed to search and destroy the infected cells. We defeat a virus by generating antibodies and cells that 'go and find these viral infected cells and kill them'. But the virus is always one step ahead. It mutates, it changes. 'So your immune system is always playing catch-up.' Some viruses mutate more rapidly than others (Hep C, HIV AIDS and influenza are particularly speedy). And as the Red Queen in *Alice Through the Looking-Glass* observed 'it takes all the running you can do to keep in the same place'.

'With viruses like HIV and Hep C we were talking about a huge amount of genetic change in literally days.' And so – stepping back from the problem for a moment – and thinking how best to combat this ever-changing enemy, Jonathan decided to approach the problem in a slightly different way.

'With viruses like HIV and Hep C we were talking about a huge amount of genetic change in literally days'

To avoid running ever faster to keep up with this ever-changing virus, he decided to try and target 'those bits of the virus that it couldn't change ... to find its Achilles heel'. The virus may be evolving rapidly, but 'it couldn't mutate and change every single bit of it' and so he launched a Europe-wide initiative to 'get data from all the genomes of all the Hep C viruses that people were collecting' and identify the bits that remained unchanged. The next step was to work out which of the many different antibodies generated by infected individuals targeted this unchanging bit of the genome. This was done by generating lots of different monoclonal antibodies (antibodies that are all clones of one specific antibody) and testing them to see which bit of the virus they targeted. 'If you've got an antibody that can latch on to these conserved parts of the virus and effectively kill them, then you've got a treatment'. And because these antibodies are 'part of your body's normal response to an infection ... you've got a safe treatment,' Jonathan explained. He was hopeful that their treatment would be suitable for use in patients who had had liver transplants and who were very difficult to treat. 'Hep C destroys the liver, and we were hoping to protect these individuals from reinfection.'

His team successfully identified 'a number of antibodies which were very good at killing several different strains of the virus in the laboratory . . . We were on the way to establishing a clinical trial to see how effective these antibodies were at killing the Hep C virus in infected humans.' But then scientists working in another laboratory found another way to stop the virus in its tracks and the antibody treatment that Jonathan and his team were working on was no longer needed. 'Unfortunately for us, but very fortunately for people infected with Hepatitis C, a number of very, very potent drugs came on the market that could eradicate Hep C, even in these difficult-to-treat liver transplant people. In terms of using antibodies to treat hepatitis C infection, there's not a great deal of interest now. It was incredibly disappointing,' Jonathan admitted. 'In terms of humanity, it was a fantastic success, and I knew I should be really, really pleased, but we'd been working on this for

'It was incredibly disappointing . . . we'd been working on this for ten, fifteen years and had been pipped to the post'

ten, fifteen years and had been pipped to the post.' When it was proved that 90 per cent of people with Hep C were cured with these wonderful new drugs, including many of the most vulnerable patients, 'it was the last nail in the coffin . . . for the idea of using antibodies to treat Hep. C'.

It was not long, however, before Jonathan's expertise felt relevant and important once again. At the beginning of 2014 an outbreak of Ebola in Eastern Guinea spread westwards to Liberia and Sierra Leone. Previous outbreaks of Ebola had happened in isolated, rural areas, and were mostly contained to at most a few hundred cases. This was on a different scale altogether. It swept through the capital of Liberia, Monrovia, and Freetown in Sierra Leone. 'Never before had the Ebola virus had so many opportunities to infect humans.' More than 28,000 cases were reported and there were in excess of 11,000 deaths.

> 'Never before had the Ebola virus had so many opportunities to infect humans'

More people died in the West Africa epidemic than were killed by all past outbreaks of Ebola put together. 'It was the largest outbreak that ever struck, and one of the things that we know is that if you take a virus from one organism and then you grow it in a different organism, very often that virus will acquire changes, mutations, that allow it to better infect that new host.' Jonathan wanted to test if this was indeed the case.

The spread of this deadly virus was a human catastrophe but it provided the perfect natural experiment for Jonathan. He wanted to understand what happens to the genome of the virus as it passed, 'from human to human to human'. 'Do any of the genetic changes give the virus an advantage in terms of its ability to infect human cells?' Does a virus like Ebola become more effective at infecting humans the more people it infects? Does a virus that's spent most of its evolutionary history living inside another host, most likely bats, get better at infecting human cells as an outbreak spreads?

His laboratory in Nottingham was not equipped to deal with a virus that killed about 40 per cent of the people it infected, so Jonathan and his team created a pseudo Ebola virus, made by sticking different surface proteins (antigens) on the surface of a similar but safer virus. They then put this Ebola-like virus in human cells and observed how it behaved, safe in the knowledge that it would not be able to replicate and that 'it would light up if there was a lot of infectivity'. In this way, Jonathan was able to show that, as the Ebola virus spreads between humans, it becomes more infectious to humans and less infectious to bats.

Suspecting that working on the front line against viruses that were killing people daily must eventually take its toll, Jim asked, concerned, 'What do you do to escape, to unwind?'

'One of the things I enjoy most is DIY,' Jonathan said, visibly relaxing as he did. 'I practise on my brothers' houses, so they're the ones with the slightly wonky walls, or the plumbing that leaks a little bit. By the time I progress to my own house, I've pretty much mastered the skill . . . The brick work is nice and straight, my plumbing doesn't leak, and my plastering is pretty good.'

The scientific foundations that have been laid, by Jonathan and thousands of other virologists around the world, have led to some incredible successes. We now have highly effective treatments for HIV and Hepatitis C. But for the individual scientists involved, success can be elusive. And at the end of a long hard week, creating something as solid and as safe as a house feels good.

KATE JONES

*Creating a global map of
disease hotspots*

Born: 1972

Grew up in: Sutton Coldfield (with her dad), Liverpool (with her older brother)

Occupation: ecologist

Job title: Professor of Ecology and Biodiversity at University College London

Viruses studied: 'global trends for all of them'

Inspiration: 'holding a tiny pipistrelle bat for the first time on the North York Moors'

Passion: 'bats'

Mission: 'to understand and protect the natural world'

Advice to young scientists: 'trust your instincts and be brave!'

PROFESSOR KATE JONES is interested in the big picture. She wants to try and spot outbreaks of disease before they occur and has created a global map of disease hotspots. As a young woman she had no interest in the study of infectious diseases and how they spread. She was much more interested in the evolution of all mammals on earth. She won the 2008 Leverhulme Prize, awarded to researchers in the early stages of their careers, for her work on the extinction risk facing many species, bats included. Working at the Earth Institute in New York, she pioneered the study of the relationship between ecology and public health, together with Peter Daszak. A decade before Covid-19, she concluded that the threat to human health from wildlife pathogens, including viruses, is 'very significant and increasing'.

Kate talked to Jim Al- Khalili in June 2015.

Two thirds of all new and emerging infectious diseases originate in wildlife. These zoonotic diseases 'are really nasty', Kate told Jim. 'Because they jumped from animals to humans quite recently, we haven't evolved to cope with them.' We are quite good at fighting familiar pathogens but, when confronted with a novel virus, our bodies struggle to launch an appropriate immune response. Viruses that are entirely new, to us and to our ancestors, can make us very ill indeed.

'Think about Ebola and HIV,' Kate said. So many outbreaks and epidemics in recent decades have come from viruses that live in animals. HIV AIDS came from chimpanzees. Ebola from fruit bats. Avian Influenza from birds. Swine Flu infected pigs. Lassa Fever is spread by rats, Zika by mosquitoes. Different species of bats seem unaffected by the coronaviruses responsible for SARS, Middle Eastern Respiratory Syndrome (MERS) and Covid-19 that they host. SARS-Cov-1 has been found in horseshoe bats and is thought to have infected civet cats. MERS-CoV infected dromedary camels. The bat coronavirus behind the Covid-19 pandemic has also been found in pangolins, long snouted ant-eating mammals.

Many scientists who study viruses zoom in on individual pathogens or epidemics and study them in great detail. Kate zooms out. She studies the grand sweep of environmental change and patterns of disease around the world. Working with Peter Daszak at the Earth Institute in New York in the early 2000s, she pioneered a new approach to the study of infectious diseases. 'We dreamt up this project of trying to understand where and when new human infectious diseases emerge.' In particular Kate and Peter wanted to know: did changes in wildlife populations and ecosystems influence outbreaks of disease? With so many viruses originating in wildlife causing outbreaks of disease in humans, it seemed highly likely that there was a relationship between ecology and public health.

'We looked at all the incidences of human infectious diseases since 1940', searching for patterns in the data and looking for possible explanations in areas that were not previously considered by scientists who study infectious diseases: patterns of land use, agricultural practices and wildlife populations. Applying her much-celebrated cocktail-making skills[1] to academia, Kate combined insights from natural and social science to try and identify the driving forces behind outbreaks of disease. And in a landmark paper published in *Nature* in 2008, she concluded that the threat to human health from animal pathogens, including viruses, was 'very significant and increasing'.

'Were you surprised that this paper was so well received?' Jim asked.

'Yes!' Kate exclaimed. 'A lot of infectious disease scientists tend to focus on particular viruses like Ebola or SARS. I hadn't appreciated that pulling everything together was actually quite novel ... And it's opened up a huge new area of macro-disease dynamics and ecology.'

---

1 After one particularly successful departmental party when she assumed responsibility for the drinks, Kate became known to many as 'Cocktail Kate'.

Kate wants to spot the threat posed by new diseases *before* they cause outbreaks: 'We need to look at how environments will change and make a forecast,' she said. A

'. . . pulling everything together was actually quite novel'

year before the Swine Flu pandemic of 2009 and more than a decade before the arrival of Covid-19, Kate created a global map of disease hotspots and showed that new infectious diseases were most likely to emerge in low-latitude countries – especially in areas where people mix freely with domesticated animals like pigs and poultry. She would like to be able to issue disease warnings, just as a weather reporter might forecast storms.

But understanding how infectious diseases emerge and spread was never on the agenda in the early part of Kate's career. As a child, she dreamed of a life spent 'in foreign parts doing exciting things', having seen Indiana Jones on the big screen, trekking through the jungle in the opening scenes of *Raiders of the Lost Ark*. 'Aged 12, I completely fell in love with Harrison Ford,' she said. 'I don't think I even took a breath until halfway through the film. It was so amazing and exciting that he was Dr Jones and, you know, I'm Kate Jones. I could see myself in his role!

'I thought maybe some combination between David Attenborough and Harrison Ford,' she said. But, attempting to reconcile her father's wishes for her to do a job with 'a secure career path and a pension', she settled on becoming a vet. Her biology teacher 'who will remain nameless' dismissed the idea out of hand. 'You're not clever enough to be a vet,' she said. So Kate thought to herself, 'Okay, I'll go and do something else with animals!'

She went to the University of Leeds to study zoology and started thinking about the diversity of life on earth. 'How are we all here? How did that happen! How can we have so many beetles?' And at the end of her second year, she spotted her opportunity for an adventure: a zoological expedition to Borneo. Sailing down rapids and searching through the forest, looking at how deforestation impacted insect populations and hunting for new species, Kate recollects feeling a lot like Indiana Jones then. On hearing of her adventures, her father suffered a 'near heart attack'. 'He wanted to keep me wrapped up in cotton wool,' Kate said. Her mother had died when she was eighteen months old and, understandably perhaps, he wanted to keep her safe.

She yearned for wild adventures in far-flung places but discovered her lifelong passion closer to home. On a university field trip to the North Yorkshire moors, a professor 'climbed a ladder and got some pipistrelle bats out of a bat box'. He put one into Kate's hands. 'They were tiny, tiny bats. If you folded one up you could probably fit it in a little matchbox. Not that you should do that! We were counting them and measuring them and then putting them back.' Pipistrelles go into a torpor in the day, lowering their body temperature and staying very still. 'And it was looking at me saying, yeah, what? What do you want? . . . I just thought it was supercool. It's not very scientific, but I just thought they were amazing!'

This cute, matchbox-sized creature introduced Kate to a group of nocturnal mammals that have been the apple of her eye ever since, despite not all bat species looking quite as cute as pipistrelles. She wrote a PhD on bat life histories, and created an evolutionary tree of all the bat species known at the time: from bumblebee bats to 'great flying foxes' and the brown long-eared bat which can detect the sound of a ladybird walking along a leaf.

'We think we know about how animals evolve,' Kate said. 'But bats break all the rules.' You'd expect animals the size of bats to live a few years, maybe about the same

'But bats break all the rules'

as a mouse, but some bats can live for up to forty years. 'They have an extraordinarily long lifespan for their size.' Free from the 'rat race' on the ground, these mammals, the only ones that can fly, have 'a completely different evolutionary strategy' and some unique capabilities. Bats can pause their pregnancies and repair their DNA.

Genealogy can be addictive. And so it was for Kate. While she was organising 1,400 species of bats, she started to wonder how all these bats were related to the other mammals in the animal kingdom. And, more ambitiously, how more than 5,000 different mammals fitted together. 'I always like big projects and a bit of a challenge,' Kate said, laughing. 'When people tell you, you can't do something, it's usually a good idea to go and do it!' she said, smiling. Being told something is 'impossible' might put some people off. For Kate it sends a clear sign to go ahead. 'Because, obviously, it's not been done before!'

Undaunted by the scale of the task ahead, she moved first to Imperial College, London and then to the University of Virginia, USA and, together with 'a bunch of really cool collaborators', started thinking about *all* the mammals on earth. How had they evolved? Where were they all now? And how likely were they to go extinct?

They collected all the published information about the genetics and physical characteristics of all the mammals that had ever lived on earth and used it to construct a set of evolutionary links, unpicking relationships that turned out not to be quite what they had once seemed, and uncovering new ones courtesy of DNA analysis.

The super tree of all mammals that Kate created was 'like a time machine'. 'You can look back at past evolutionary events and look at what prompted diversification or how species have gone extinct or have evolved into different things . . . It helped provide a framework within which to answer evolutionary questions.' And it's a way to find out more about why some species of mammal are more endangered than others. The 'flip side' of thinking about how all the mammals on earth came to be here is thinking about how we can stop them from disappearing. 'How can we protect the biodiversity of the planet?'

Kate moved back to the UK in 2005 when her father became ill, and got a job at the Zoological Society of London. She was determined to do something to stop the dramatic loss of species that was occurring during her lifetime. And, before long, she had found a way to fulfil her childhood ambition and act like her namesake, Dr Jones, once again.

'Basically, I started a global bat monitoring project in Transylvania,' Kate said, laughing. Romania is one of the most species-rich areas for bats in Europe and so it seemed like a good place to study whether or not bat populations were declining. Curiously, given all the gothic horror stories, the only bats she couldn't monitor were vampire bats. There are only three species of vampire bats. None of them live in Transylvania and they don't suck blood. They make a razor-sharp incision in their victims – typically cattle not humans – and their saliva contains substances known as draculin that stops the blood from clotting. The blood flows freely and they lap it up just like a kitten would a bowl of milk. Bats, in general, are far more likely to *eat* bloodsuckers like mosquitoes than to swoop down, Dracula-style, and bite your neck for their next meal.

They do, however, have an enormous appetite for insects. Bats pollinate plants and disperse seeds. Bananas, mangoes, the blue agave plant (the base ingredient in tequila) and many more plant species depend on bats for their survival. They are sensitive to changes in their environment and live across a wide range of different habitats, from woodlands to urban areas. Kate figured that trends in bat populations ought to be a good indicator of what was happening to biodiversity more generally and so she decided to launch a project to track and survey the bats using acoustic monitoring.

'Bats leak information about themselves into the environment,' Kate explained. By recording all their chit-chat, courting songs and echolocating activities, she hoped to build up a picture of all the different bat populations in Romania and how their numbers were changing over time. Acoustic monitoring on the scale that Kate was proposing was unheard of when she first suggested it to colleagues. She remembers many of them saying: 'You can't possibly do this.' 'But, you know, it's like a challenge,' Kate said. 'That was a red flag to a bull really!' From that moment on, it was of paramount importance to Kate to prove that her colleagues were wrong to doubt her.

'We got together with our partners in Romania, the Romanian Bat Protection Association,' Kate said, 'offering training and donating equipment to them.' They put microphones on top of cars and started 'driving around Romanian roads, gathering acoustic data'. Within a few years, citizen scientists all over Romania and in Bulgaria, Hungary, Ukraine, Russia, Mongolia and Japan, were chasing bats with microphones. And Kate had invented iBats, an app which enabled anyone with a mobile phone to record bat sounds, and automatically linked them to the location where they were heard. As more bat detectives sign up to the project and more time passes, the database is growing. Bat calls recorded across Eastern Europe over a ten-year period should be enough to uncover statistically significant trends in bat populations and give us a good indication of how the overall biodiversity in these areas is changing over time.

When Kate started her bat monitoring project, her colleague Peter Daszak was sceptical. 'Why do you want to do that?' he asked. But pursuing her desire to preserve nature in all its richness, she has contributed to the study of disease. Changes in biodiversity can have a significant impact on global health. When areas that were once species-rich – rainforest for example – are cleared for human use, it becomes much more likely that a virus found in wildlife will have the opportunity to infect us, or our domesticated animals.

'Nipah is interesting,' Kate said. 'Because we think its emergence has a lot to do with anthropogenic change.' This virus was first identified in humans during an outbreak on a pig farm in Nipah, Malaysia in 1999. There were 265 pig farmers suffering from fever, vomiting and general confusion and 105 of them died, many with very swollen brains. Initially it was misdiagnosed as Japanese encephalitis. The Nipah virus, as it soon became known, looked similar to the deadly Ebola virus and, like Ebola, induced a haemorrhagic fever. But while Ebola attacked blood vessels, causing organs to fail, Nipah attacked the brains of its victims, inducing a coma.

> 'But while Ebola attacked blood vessels, causing organs to fail, Nipah attacked the brains of its victims, inducing a coma'

With more people living in and around the town of Nipah, small-scale pig farms were being replaced by larger commercial pig farms. As the pig farmers expanded their grazing sites into the forests, they began to occupy areas inhabited by fruit bats. With the bats playing host to the Nipah virus, the first crossover occurred when the pigs started to eat fruits lying on the forest floor which had been cast off by infected bats. Droppings and urine from the bats also found their way into the pigpens. Once the virus had infected pigs, it was only a matter of time before humans also caught the disease, either through direct contact with sick pigs or their contaminated meat.

Detailed location-based case studies like this can help us to understand how outbreaks happen and could help to provide practical advice on outbreak prevention. The modelling techniques Kate developed to study Ebola outbreaks in the Democratic Republic of the Congo accurately predicted an outbreak of Ebola that occurred several years later. Her models of Lassa fever in Nigeria revealed how outbreaks of this disease would follow like clockwork three months after heavy rains, a trend that Kate was then able to explain as follows: rat populations go through cycles of boom and bust, depending on the amount of rain. More rats are conceived during the wettest weeks and when they reach three months of age, they start to spread the disease.

There are no hard-and-fast rules, but the spread of infectious diseases is closely linked to what's happening in wildlife populations and ecosystems all around us. Kate loves nature for nature's sake. 'I want to live in a world where there are tigers,' she says. And 1,400 different species of bats. 'It's an enormous challenge, balancing the needs of spiralling global populations, the needs of land for food and for making a living, against the needs of wildlife ... Is it possible to meet all those needs without some serious sacrifices somewhere?' she asked. Then answered her own question: 'I think we need to try.'

Preserving the richness and wonders of nature is a worthwhile endeavour in its own right. It could also help to stop animal viruses from spilling over into humans and our domesticated animals. The viruses that live in the wildlife all around us are not going to go away. Most of these wildlife viruses have lived in their animal hosts for hundreds of thousands of years without causing us, or our animals, any harm. How we use the land, where we live and how we farm can all have an impact on what these viruses might do next.

'But perhaps the best way to inoculate the planet against the next pandemic is to respect nature and, as far as is possible, keep wild places wild'

If we are infected by a novel virus, we can search for vaccines after the event. But perhaps the best way to inoculate the planet against the next pandemic is to respect nature and, as far as is possible, keep wild places wild. We are part of nature. We are in it, not above it, and when we forget that we are just one of many millions of species on this planet, there can be serious consequences for animal and human health.

MARTHA CLOKIE

*Hunting for viruses
that cure not kill*

Born: 1973

Grew up in: the Scottish Highlands and Islands

Occupation: microbiologist

Job title: Professor of Microbiology at the University of Leicester

Viruses studied: the viruses that eat bacteria

Inspiration: 'the realisation that viruses could cure disease'

Passion: 'finding out what viruses are doing in nature'

Mission: 'to create virus cocktails that can be used as medicines'

Advice to young scientists: 'follow what you find exciting'

PROFESSOR MARTHA CLOKIE hunts for viruses that infect bacteria and uses them to develop medicines. She worked as an evolutionary biologist and ecologist for many years, studying the evolution of the African violet in Uganda and the viruses associated with the cyanobacteria that live in the ocean. In her thirties, she met some scientists from Georgia who were using viruses to treat wounds and, excited by the possibility of using viruses to cure disease, she decided to become a medical researcher, swapping the study of pretty plants for staring at stool samples under a microscope. For seven years she searched for the viruses she needed to treat a particularly nasty form of antibiotic-resistant diarrhoea. Eventually, she found what she was looking for buried in a salt marsh.

Martha talked to Jim Al-Khalili in October 2019.

There are trillions of viruses on this planet. 'The numbers are truly astronomical,' Martha told Jim. In fact, there are more viruses in the biosphere than there are stars in the visible universe: 10 million times more. Or put another way: if all the tiny viruses in the biosphere were lined up end to end they would extend to the edge of the visible universe and back.

'Every living thing has got its own set of viruses,' Martha told Jim: dogs, cats, kangaroos, and cockatoos; tobacco and tomato plants, fruit flies, fungi, bumblebees and bacteria. Viruses are the most abundant organisms on Earth and not all of them are out to get us. 'The relationship between a virus and its host is very specific' and most of the viruses on this planet infect bacteria, not us. These are the viruses that Martha wants to find. The viruses that have the potential to cure disease.

As an undergraduate at Dundee University studying botany and genetics, she was more interested in 'the things she could see – the visible world'. 'I did a little bit of microbiology but it was not my favourite subject back then,' she told Jim.

Botany was her first love. A passion for plants was in her blood, inherited from both her parents and her grandmother before that. 'My granny was a brilliant botanist. She read French at Oxford but sorted out all the specimens in the Oxford Herbaria and wrote the definitive guide to the collection that is still in use today.' Martha's mother ran a plant nursery and her father 'was an environmental consultant for Oilfield in the Highlands of Scotland ... and later had a business selling sea vegetables (seaweed), mainly for the macrobiotic and the high-end food market'. As a result of both of those activities, Martha and her three brothers were taken 'every Saturday, and most Sundays' to the beaches where they lived, forty miles north of Inverness. 'We went all the way up to the top of the UK and across to the west coast as far south as Skye,' she said. 'Essentially my father would leave us at the top of the beach while he went to go and do his surveying and get his seaweeds. And being left on those beaches ... with not a huge amount else to do, I would figure out what all the seaweeds and animals were in the rock pools, what all the plants were at the top of the beach.'

Later, her specialist subject was African violets: the evolution of the twenty different species which 'grow all over Kenya and down into Tanzania'. 'I really loved the fact that you could look at the genes of this plant to see how long it had taken for these plants to evolve. And work out when one species becomes another. Who is more related to who?' Looking at the different flower and leaf shapes and parts of genomes from plants and studying how they had changed over evolutionary time was interesting but painfully slow work. In an attempt to speed things up a bit, Martha decided to spend some time studying 'something a little bit simpler': photosynthesising bacteria (called cyanobacteria) 'that divided every day instead of a minimum of every six weeks'.

'You can think of cyanobacteria as primitive plants.' They are also known as blue-green algae and it's the chlorophyll and other pigments in these photosynthesising bacteria that gives them their distinctive colour. Understanding how a single-celled bacteria evolved seemed like a much more manageable proposition than getting to grips with an entire plant. 'You could have many different populations of these bacteria in small flasks and do more experiments on them and over much shorter timescales.'

Martha wanted to work out what was driving the diversification of this cyanobacteria, which have about twenty-five different sub-types. To do this she explored all sorts of different evolutionary pressures. The effect of acidity and temperature on the bacteria in flasks, for example. Her boss, Nick Mann, who was the Principle Investigator on the cyanobacteria research project had another idea.

'In 1989, just two or three years before I started work-ing on the project, scientists had made an astonishing discovery.' As well as there being extremely large numbers of the cyanobacteria in the oceans, there were even larger numbers of these tiny particles, which when viewed under an electron microscope turned out to be viruses. The ocean, which was previously thought to be virus-free, was found to be teeming with these tiny particles. No one was expecting that.

'A significant chunk of the viruses on planet earth are in the oceans and in the ocean sediments,' Martha said. Viruses are known for their ability to shake things up genetically. They invade the DNA of their host, often breaking it up or sometimes adding new pieces. These DNA fragments then join together forming novel com-binations of genetic 'letters' and so can introduce variety within a species. Nick was one of a few scientists at the time who thought these newly discovered ocean viruses might be the driving force behind the diversification of cyanobacteria and he asked Martha to investigate them as well 'in case she might find some answers there'.

He wanted her to see if there was a relationship between the viruses that you find naturally occurring in those waters and the diversity of those bacteria. Martha had got this far in her career without giving viruses a second thought, but no matter. She did as she was asked and set off to sea in search of the ocean viruses that infected cyanobacteria and it turned out to be a good move.

'How easy was it to find these viruses in the ocean?' Jim asked, thinking that finding a particle that was a hundred thousand times smaller than the width of a fingernail might be challenging.

'It was quite easy to find them,' Martha said. 'I knew how to bait them, as it were . . . I could go to more or less any bit of seawater and find these viruses.' Answering the question she had been set by Nick was more tricky. To do this, she needed to understand how these ocean viruses were able to penetrate their bacterial hosts and this meant she needed to study their tails. Viruses use their tails to attach themselves to the bacteria, then create a tunnel through which they inject their genetic material into host cells. It's the tails that make the relationship with the bacteria they infect so 'exquisitely specific'. However, try as she might, Martha couldn't find these tails based on all known viruses.

'So I decided that the best approach was to actually sequence the genome of one of these viruses to see if I could understand some of its features,' she said. Such an approach would have been impossible a few years earlier, but new sequencing-based techniques to study whole genomes were now more affordable. And this meant it was possible, in theory at least, to read every letter of the genetic code on a piece of DNA. Sequencing the genomes in this way opened up the entire instruction manual of an organism. Previously, the best that scientists could do was to search for individual genes, and they would miss any novel and potentially more interesting information.

'It was really enormously exciting, at the time, to sequence a whole genome of something,' Martha said. 'Especially as these viruses are not trivial, not tiny ...' Typically their genomes are about 200,000 genetic letters long and contain 200 different genes. 'And we found out all sorts of things that really – in our wildest dreams – we hadn't expected to find.' Many of the genes Martha found had already been found in other viruses. This was expected. But there was also 'loads of stuff that we didn't know', Martha said. 'Intriguingly, they had grabbed lots of things from their bacterial hosts, including the key genes used for photosynthesis!'

'And we found out all sorts of things that really – in our wildest dreams – we hadn't expected to find'

It took Martha, and scientists elsewhere in the world, several years to work out what was going on. Eventually they were able to explain this highly unexpected finding: when a virus infects a cyanobacteria, it doesn't destroy the cell that it infects. It keeps it 'just ticking over, like "you will make the energy I need"', Martha said in her best Dalek voice. 'So it is a bit like a sort of slightly sinister life-support machine, because it is a viral-support machine.'

'So the virus keeps the bacterial cells alive for just long enough to make more viruses!' Jim exclaimed. 'That's so sneaky!'

Previously it was thought that these viruses – bacteri-ophages – destroyed their hosts. Phage is a Greek word meaning 'to devour'. But studying their DNA revealed just how sophisticated these apparently simple organisms could be and showed that the relationship between ocean viruses and their bacterial hosts was much more interesting and intricate than had anyone had imagined. 'It's not a simple infect and kill.'

'It's not a simple infect and kill'

At this point, it was clear to Martha that if she wanted to understand cyanobacteria, she could not afford to ignore the bacteriophages that lived inside them.

'Because I had come to phages fairly late in life ... I had to quickly try and skill myself, as best I could. So I read everything I could find, every paper, every book, and then I tried to track down some conferences on phages.' The first phage conference that Martha attended, 'in a place called Olympia, not far from Seattle', changed the course of her career.

She arrived a couple of days early, excited to be travelling to the West Coast, and stayed with the conference organiser, who put her to work 'chopping salads and editing posters'. The posters prepared by teams of scientists from Georgia were a revelation. 'There were lots of case studies of patients with horribly infected wounds that had been cleared up by phages,' Martha said. 'That sort of blew me away! I had no idea you could use a virus to treat an infection at that point.'

The next day, Martha displayed her own poster at the conference. Jim wondered how the phage community felt about an outsider gatecrashing their conference. 'I had never been received so warmly in any meeting ever,' Martha said. Scientists who worked on other phages

stopped to look at her poster and 'were excited to discover their genes in something else'. They generously
shared their knowledge of the genes they had found, for
example in bacteriophages associated with E. coli, and
helped Martha to fill in gaps in her knowledge about the
marine bacteriophage she had sequenced. But it was the
idea that viruses could be used to tackle disease which
stuck in Martha's mind. 'It was part of my personal "Oh
My Goodness" moment,' she said. And it led Martha to
embrace what was then a deeply unfashionable area of
medical research.

'Why were you keen to get involved in an area that I think
many researchers would have considered to be a bit of a
dead end?' Jim asked.

'Well, it just seemed to me a logical thing to try and
look at,' Martha said. 'There was hard evidence to show
that bacteria were becoming resistant to antibiotics.'
And there was a desperate need for new treatments for
some old diseases. 'Where are we going to get the new
antimicrobials?'

Before the Second World War, phage treatments were
not uncommon for all sorts of gut and skin infections. But
with the arrival of miracle cures like penicillin, interest
in this highly targeted way of treating disease nearly dis-
appeared. Antibiotics were the future. Phage treatments
were banned in the Soviet Union for being 'backward'. But
distrustful of diktats from the Soviet Union, many house-
wives, particularly in Georgia, kept hold of these tried
and trusted remedies for as long as they could. Collections
of the phages used for different ailments were maintained
by the staff at the Eliava Institute of Bacteriophages,

Microbiology and Virology who kept them safe in their fridges at home during turbulent times.

Outside of Georgia, there were 'a few little pockets of interest' in using bacteriophages to treat disease, 'in Russia, France and Poland'. But elsewhere in the world, research in this area had come to a grinding halt. Bacteriophages, which targeted just one strain of a particular bacteria, were thought to be of far less use than broad-spectrum antibiotics, like penicillin, which cured all manner of ills. 'One antibiotic will often work on all the different members of a species. If it kills one strain of a bacteria, chances are it will kill them all. Not so with phages.'

As more bacteria developed antibiotic-resistant strains, the very thing that had once made phages seem inferior made them look attractive once again. 'You have got these viruses that have exquisitely evolved to kill specific bacteria, so it seemed to me obvious to try to look at them.'

Martha also thought it would be a good idea to try and bridge the gap between two 'very different communities of people' who were studying phages in the 2000s: the medical researchers who searched in sewage for the viruses associated with the bacteria that caused human diseases, who then purified and tried to use the viruses directly; and environmental phage scientists like Martha, who were looking for viruses in the soil and in the ocean and wanted to understand what viruses were doing in the natural environment. 'I wondered if I could use those things that I had learned from studying these environmental phages and apply them in this different field,' she said.

Her botanist parents thought her desire to move into medicine was 'a bit odd'. 'They were moderately supportive,' Martha said. 'But I don't think at the time they thought it was a particularly good idea.'

Fortunately, a promising job came up at the University of Leicester, where Martha had studied for her PhD. 'They wanted somebody who could teach environmental micro-biology and work in a medical school, and I just thought: "Oh, I need to get that!"' Martha said.

The next step was to convince the medics that she had something to offer them. Face to face with several very senior academics in a 'quite formal, wood-panelled room', Martha was asked: 'Why should we give you, a botanist, a lectureship in a medical school?' What use was a plant specialist to medicine? The evolution of African violets or cyanobacteria in the ocean was interesting, but hardly a reliable substitute for seven or more years of medical training. Thinking fast, Martha made the connection for them. 'I told them that, in my opinion, the human body was just an interesting interconnected set of ecological niches. And if we viewed the human body more like that then perhaps, we would be able to better understand dis-ease.' Hearing this, they warmed up slightly. 'I could see that they thought that was slightly interesting,' Martha said, wryly. And, remembering that two of the most senior scientists on the panel were eminent cardiologists, she told them about the phage that causes endocarditis. Doctors are taught that this condition occurs when blood vessels become infected by the bacteria, Streptococcus mitis. But Martha told the medics: 'Actually, the thing that causes the disease is a bacteriophage that hides inside this bacte-ria, it makes the bacteria extremely sticky and they block the heart valves.'

'They definitely stopped looking bored at this point!' Martha said, laughing.

She got the job and was told: 'You can do what you like really as long as you write good papers and bring in grant income and teach well, so that was very freeing.'

Looking around for medically relevant pathogens to research, Martha found 'there was a fair amount of work being done on phages associated with Pseudomonas, E. coli and the multi-drug resistant Staphylococcus aureus (MRSA).' But one superbug seemed to have been over-looked. 'In Leicester, and actually in England in general, at the time, we had really, really high rates of Clostridium difficile,' Martha said.

'It is a complicated bacteria,' she explained. 'People often haven't heard of it, because it generally infects old people when they go to hospital with some other con-dition, like a chest infection.' Sadly, the antibiotics they are given to treat their original infection can create other problems. They 'wipe out the whole of the gut flora' in-cluding the good bacteria in our guts that help to keep C. difficile in check. Once C. difficile starts to proliferate, it's hard to stop, 'like a weed, growing wild in the gut'. It is 'naturally resistant to nearly every antibiotic' and notori-ously difficult to treat.

When C. difficile takes hold, 'it causes a sort of leaky junction in gut cells, resulting in this horribly infectious and nasty diarrhoea'. It is a very serious infection. For about one in eleven patients, it is fatal. 'The British gov-ernment, particularly, has been really good at controlling it by having improved hand hygiene and hospital hygiene, but it is still there, and it is still killing tens of thousands of people worldwide a year,' Martha said. 'In places where they haven't got that top-down control, like the United States, it is a particularly important problem. And it has not gone away.'

'Perhaps it is worth mentioning here that C. difficile is so named for good reason,' Jim said.

'I didn't really think about that at the beginning!' Martha said, laughing. 'I just had, well, a youthful exuberance. I could see that there was very little done on the C. difficile viruses and I just thought: well, it is probably because they haven't done it right! I am going to do it right ... and I can do it!'

> *'I could see that there was very little done on the C. difficile viruses and I just thought: well, it is probably because they haven't done it right'*

Spurred on by her success with the cyanobacterial viruses 'which are really, really hard', Martha assumed that, despite their reputation for being difficult to study, she would somehow manage. She wrote a grant to the Medical Research Council proposing that she would try to identify a set of viruses that could perhaps be used to develop both new treatments and new diagnostics for C. difficile. 'Naively I remember I had a planning chart which said: 0–6 months – find viruses.' As things turned out, '0–6 years' would have been more appropriate.

> *'I remember I had a planning chart which said: 0–6 months – find viruses'*

'So how did you go about looking for these viruses?' Jim asked.

'Well, to begin with, I just followed my own dogma, and the dogma in the literature, which is: wherever you find bacteria, you will find viruses ... For every bacterial cell, there are between ten and 100 viruses ... I developed a nice relationship with our infectious disease doctors and clinical microbiologists at the hospital and, after lots of complicated ethics and so on, we got given the leftover stool samples from patients who were suffering from chronic diarrhoea.'

'. . . wherever you find bacteria, you will find viruses . . .'

The work was not particularly pleasant, 'but it was a logical place to start'. 'They were extremely smelly, disgusting samples.' But, reluctant to delegate work she wasn't prepared to do herself, Martha put a lab coat on and got stuck in, looking for viruses in stool samples with the other members of her team. For a year and a half, Martha and the team got only negative results. 'We screened hundreds and hundreds of samples, sadly to no avail. We just didn't find a single virus associated with those C. difficile bacteria.'

'Were you tempted to give up?' Jim asked. 'Maybe C. difficile was too difficult?'

'I just thought: well, they must be somewhere,' Martha said, matter-of-fact.

Winning her first major research grant when she was pregnant with her first child was not ideal; but there were some advantages. A study of the literature on C. difficile bacteria revealed that it was more abundant in the guts of children than those of infected adults. And since Martha was spending a lot of her non-work time changing nappies, she decided to look there as well.

'The sampling wasn't quite as disgusting.' And, screening the contents of her son's dirty nappies led to an exciting result. 'I found a good virus in my first son Rowan,' Martha told Jim. 'We called it R9, because he was nine months old at the time.' Keen to replicate

> 'I found a good virus in my first son Rowan'

these findings in case her son was the exception not the rule, she scooped up whatever dirty nappies she could find. 'I remember going to these baby drop-in sessions that you are supposed to go to when you become a new mum and I would just round up all the nappies from all the babies there. I said, "Don't get rid of those!"' Until one mum, a medical ethicist, objected, reminding Martha that poo is covered by the Human Tissue Act and so could not be used for medical research without the owner's consent. As for the other mothers, 'I think they thought I was an idiot,' Martha said. 'You are just supposed to be in love with your baby and nothing else. I was very much in love with my baby, but I didn't want to waste the opportunity.' And, so with a little modification of procedure to accommodate ethics, she transported bags full of dirty nappies from the

> 'I was very much in love with my baby, but I didn't want to waste the opportunity'

baby drop-in centre to her laboratory for screening.

'It actually looked like babies were the answer,' Martha told Jim. But sadly, the discovery of R9 turned out to be a bit of a false dawn. 'There are several hundred different types of C. difficile, what we call ribotypes, and some are worse than others for humans.' For viruses to be effective treatments, they need to be able to kill the strains of a bacteria that are clinically a problem. The relationship between phages and the bacteria they infect can be 'exquisitely specific'. 'One of the reasons that phage therapy became unpopular is because phages are very specific, not just to the bacteria but to the subtypes of bacteria,' Martha explained.

Armed with this knowledge, Martha had 'built up a nice set of strains that she knew were problematic in the UK and Europe', ready to test the effectiveness of the viruses she was hoping to find. To her great disappointment, the viruses she had found in the babies' nappies 'weren't brilliant'. They only killed a few strains of C. difficile, none of which caused particular problems in humans.

The search continued and, being an ecologist at heart, Martha looked beyond hospitals and patients, to nature, for inspiration. She travelled round the country, collecting mud and estuarine samples and storing them, carefully labelled, in a fridge in the back of her car. By this time, her sons Rowan and Alasdair were in wellies, not nappies, and were 'quite keen sampling assistants'. Many a weekend was spent on family virus-hunting trips in search of the best kind of mud. 'I tested different layers of the mud,' Martha said. 'The most successful was the layer just below the surface layer where the mud becomes black and free from oxygen.'

Seven years after her search had begun, Martha finally found the viruses she was looking for, buried in a salt marsh in a muddy estuary in Hampshire. When she tested these samples in the lab, they gobbled up the strains of C.

difficile that were making us ill. 'So, we were jumping up and down with excitement at this point,' she said.

In a landmark paper in 2006, Martha described how the phages that she had first found in the thick, black, anoxic Hampshire mud could be used to prevent hamsters from becoming infected with C. difficile. 'It's all very well to show that something works in a flask, but we wanted to use it on patients,' Martha said. 'You can't study new phage-based treatments in acute C. difficile infections in humans, as it would be unethical', but pigs and hamsters are a good proxy. And so Martha asked her good friend Gill Douce if she could help. Gill had spent five years developing a hamster model to study C. difficile bacteria and got it working 'really nicely'. 'I asked her if she could possibly test some of my phages. And she said no! She was doing lots of vaccine work and she wasn't convinced that phages would be very effective.

'But then one of her team, Janice Spencer, had previously worked on phages in disease. So we just sort of battered poor Gill until she agreed to put some phages in her hamster model and she was really surprised and pleased at how well they worked!' There was a lot of uncertainty around these experiments and absolutely no guarantees that they would be a success. But, to Martha and the team's delight, 'the phages really significantly reduced the amount of C. difficile that could grow in the hamsters' guts'. In this way they showed that hamsters with these phages in their system would be less likely to succumb to an outbreak of C. difficile infection. Even when they were exposed to 'lots of C. difficile … the C. difficile won't grow because the virus kills it straight away'. This work was published a few years later in a seminal paper that described which types of C. difficile the phages could kill alongside the data on the phages themselves and their efficacy in flasks and hamsters.

Prevention is better than cure. And it makes even more sense when you're trying to treat a hospital-acquired infection like C. difficile. 'You can just imagine it becoming a standard practice for people going into hospital to be given some kind of drink or capsule containing a cocktail of the viruses that target C. difficile,' Martha said.

All of Martha's subsequent experiments – and there have been plenty – support her original findings. 'We have a good clear set of targeted viruses that only kill C. difficile . . . and all of our data suggests that if we can get these viruses into humans before they are infected, we will be able to prevent infection.' Funding for clinical trials has not been forthcoming, but Martha is hopeful that one day her therapeutic viral cocktails could be used to prevent and treat this difficult disease, and is continuing her work with these phages to ensure that they can be developed in a safe and effective way.

For decades, the medical establishment in the Western world was highly sceptical of anyone who thought viruses could be used to improve our health. Even as recently as 2006, Martha was warned by fellow phage scientists to avoid making any mention of medical applications in grant proposals. They knew, from experience, that to do so would ruin Martha's chances of success. Scientists who expressed an interest in phage medicines were generally considered to be either desperately out of date or just plain wacko. Today, finding alternative treatments for a growing number of diseases which are (or soon will be) antibiotic resistant, is the holy grail. And trying to find phage medicine to defeat superbugs is a highly active field of medical research.

Phage treatments are being developed for some of the most serious infections in the UK, including MRSA, Pseudomonas and sepsis. Martha's lab is looking at phage treatments for urinary tract infections. 'UTIs would be a very, very, good target for phage therapy,' she said. As many sufferers will know, the bacteria that cause urinary tract infections can be hard to treat. This is partly because 'the bladder is a hard place for an antibiotic to get to'. And there is another good reason to focus on treatment for UTIs. 'About half of all sepsis cases are thought to be caused by urinary tract infections.' An effective treatment for UTIs could radically reduce the number of people who die from sepsis.

Phage research in Europe and the USA tends to focus on improving human health. 'In Asia, there's a lot of interest in treating animals with phages . . . If fewer antibiotics are used to treat animals it will help prevent that build-up of resistance being transferred to humans.' Many countries, including China 'now have limits on the amount of anti-biotics that they can use in animal production'.

Aware of this, and a little frustrated by the logistical and regulatory hurdles she had to jump to initiate human clinical trials, Martha thought research on pigs might be more straightforward. Salmonella is a major problem in pigs, as well as chickens. And so she decided to look for the viruses that could help to stop pigs from getting infected.

The seven years Martha had spent hunting for viruses associated with C. difficile helped to inform her strategy.

From the start she searched in healthy individuals as well as sick ones, having reached the counter-intuitive conclusion before that sick humans tend to be phage-depleted. This is because the presence of these phages seems to keep the harmful bacteria in check. Healthy people have potentially harmful bacteria in their guts. It's only when they proliferate that these bacteria cause infections. This occurs when there are no viruses to keep their numbers down. The viruses in our guts help to maintain a healthy balance of many different bacteria in our microbiome. Similarly, the viruses in the ocean keep cyanobacteria in check and help to prevent algal bloom.

When Martha told the Leicester medics that it might be fruitful to think of the human body as an ecosystem, she knew none of this. It seemed totally logical based on everything she knew, as well as being a hunch conjured up on the spur of the moment to impress some eminent professors and secure the job of her dreams, and has proved to be a useful way of understanding disease.

Hunting for C. difficile viruses was a lot less glamorous than studying African violets. And her next research project was equally unpretty. She found the salmonella viruses she wanted in pig shit and the mud surrounding pig styes. Her team found these viruses quite quickly, but it took three years to characterise them and show that they were effective on all the relevant forms of Salmonella that cause disease in UK pigs. Fortunately, they were heat-resistant, which made it a lot easier to incorporate them, in a dried form, into pig feed.

To test the effectiveness of her treatment she worked with vets in Nottingham and set up an experiment in which 50 per cent of the piglets in her study were given their normal food and 50 per cent were given phage-enhanced pellets. The results were clear. Phage-fed pigs were much less likely to get sick from salmonella. This pilot study helped Martha to secure a generous grant from the British Biotechnology and Biological Science Research Council. If pigs could be treated with phages instead of antibiotic it could lead to a significant reduction in our use of antibiotics.

'There is a really pressing need to reduce the use of antibiotics,' Martha said. 'We've predicted that unless something is done about our chronic over-reliance on antibiotics to treat both animals and humans, approximately 10 million people a year are going to die from antibiotic-resistant infections worldwide.'

But first the phages associated with the bacteria that make us ill need to be located and understood. The viruses Martha needed to treat C. difficile were buried in a salt marsh; she found the necessary pig salmonella viruses by chasing down wild boars in the Forest of Dean. 'I have a really interesting project at the moment,' Martha said, 'looking for phages in some of the oldest rainforest in the world, in Malaysia.' If we want to find new viruses, it makes sense to look in places that are species-rich. We know there are millions of viruses out there that keep harmful bacteria under control, some of which could help treat our infectious diseases. Finding them will require the skill and ingenuity of some committed virus hunters.

Viruses that infect humans can cause epidemics and pandemics. Viruses that eat bacteria (bacteriophages) could be the medicines of the future.

ACKNOWLEDGEMENTS

Thank you, first and foremost, to all the scientists featured in this volume. These are your stories. I am deeply grateful to you all for being so generous with your thoughts and for being so honest about the things that didn't go so well, as well as those that did. Not to mention helping me finalise these chapters when you had many more important things to do.

Gwyneth Williams commissioned *The Life Scientific*. It was her brilliant idea to broadcast in-depth conversations with scientists on BBC Radio 4 and it was embraced whole-heartedly by the current controller of Radio 4, Mohit Bakaya. Thank you both for creating such a wonderful opportunity for the BBC Radio Science Unit. Alan Samson at Weidenfeld & Nicolson commissioned this series of books and believed in me as an author. Thank you, Alan. And thank you Maddy Price for your editorial insight and Clarissa Sutherland for keeping everything on track! Madeleine Finlay wrote some wonderfully witty copy for the chapter on Kate Jones. Naomi Darlington, Eliza Quint and Rosie Quint transcribed the interviews much better than a computer.

I am very lucky to have such great colleagues in the Science Unit and special thanks are due to: Andrew Luck-Baker (who produced the interview with Jeremy Farrar),

Deborah Cohen (Wendy Barclay), Beth Eastwood (Jonathan Ball) and Michelle Martin (Kate Jones). Thank you for some great interviews that were a joy to turn into stories. And thanks to the editor of the BBC Radio Science Unit, Deborah Cohen, in particular. I have learnt so much from you over the years. Our fabulous production co-ordinator, Maria Simons keeps the show, and me, on the road. And I am, of course, deeply indebted to Jim Al-Khalili. It has been a pleasure and a privilege working with you over the years, trying to make every programme just as good as we possibly can. We greatly appreciate your intelligence, humility and extraordinary ability to make guests feel at ease.

Thank you to my brother, Will Buckley, for his positivity and encouragement and my friend Kate Steiner who provided moral support and wise advice when it was most needed. Writing books is a time-consuming process and I am profoundly grateful to my husband, Mike Quint, and our daughters Eliza and Rosie for putting up with all the hours that I haven't been there. Finally, thank you Mike for keeping my eyes open to alternative perspectives and for understanding me perhaps better than I understand myself.

INDEX